MATHEMATICS:PROBLEMS AND SOLUTION CLASS 9TH NCERT BASED

Dr. Lal Chandra

GOVERNMENT HIGH SCHOOL HALIYA MIRZAPUR

Email. lal9026153385@gmail.com

October 12, 2024

Acknowledgements

Present book comes in existence to solve the dificulty of student class 9th of the subject mathematics. The book contains all the examples of NCERT mathematics book and their exercise. The solution of examples and exercise in simple and logical form that faciliate the students to understand the concepts and steps of solutions with proper framework. The study of this book will make more skilled to derived the answer of problems to reader accurately and efficiently.

Contents

Acknowledgements	ii
Contents	iii

1 NUMBER SYSTEM — 1

- 1.1 Introduction 1
- 1.2 Rational numbers 3
- 1.3 Solutin of Exercise 1.1 6
- 1.4 Irrational Numbers 8
- 1.5 Solution of Exercise 1.2 10
- 1.6 Real Number and Decimal Expantions 11
- 1.7 Solution of Exercise 1.3 16
- 1.8 Operations and Real Numbers 25
- 1.9 Solution of Exercise 1.4 31
- 1.10 Law of Exponent for Real Numbers 34
- 1.11 Law of Extended Exponents for Real Numbers 35
- 1.12 Solution of Exercise 1.5 37

2 POLYNOMIALS — 40

- 2.1 Introduction 40

CONTENTS

2.2	Polynomials in One Variable	40
2.3	Type of polynomial	42
2.4	Solution of Exercise 2.1	44
2.5	Zeros of a Polynomial	48
2.6	Solution of Exercise 2.2	49
2.7	Factorisation of Polynomials	55
2.8	Solution of exercise 2.3	57
2.9	Solution of Exercise 2.4	61
2.10	Algebraic Identities	68
2.11	Solution of Exercise 2.5	74

3 COORDINATE GEOMETRY — 89

3.1	Introduction	89
3.2	Solution of Exercise 3.1	90
3.3	Cartesian System	93
3.4	Solution of Exercise 3.2	101

4 LINEAR EQUATIONS IN TWO VARIABLES — 105

4.1	Introduction	105
4.2	Linear Equations	105
4.3	Solution of Exercise 4.1	107

CONTENTS v

    4.4   Solution of a Linear Equation 109

    4.5   Solution of Exercise 4.2 110

5 INTRODUCTION TO EUCLID'S GEOMETRY   114

    5.1   Introduction: 114

    5.2   Euclid's Definitions, Axioms and Postulates 115

    5.3   Solution of exercise 5.1 122

6 LINES AND ANGLES   127

    6.1   Introduction 127

    6.2   Basic Terms and Definitions and Notations 127

    6.3   Solution of Exercise 6.1 137

    6.4   Lines Parallel to the Same Line 144

    6.5   Solution of Exercise 6.2 149

7 TRIANGLES   156

    7.1   Introduction 156

    7.2   Congruence of Triangles 156

    7.3   Criteria for Congruence of Triangles 158

    7.4   Solution of Exercise 7.1 164

    7.5   Some Properties os a Triangle 175

    7.6   Solution of Exercise 7.2 181

CONTENTS vi

7.7 Some Other criteria for Congruence of Triangle 190

1 NUMBER SYSTEM

1.1 Introduction

Story of number: When Yash wlaking at the bank of river, he collected the white stones. During the collection Yash tells, the number of stones to his sisther Lalita as; 1, 2, 3, 4, ... This number which represent the stone of Yash that collected by him is called Natural number and it is denoted by N. Now when he transfer all stones to his sister then the stone left to his is represent by symble 0 and called zero. The numbers which need to Yash to expressed stone is called whole number which denoted by W. Now Yash gave all stone to Lalita but Lalita is demanding one more stone because he was prommised to Laltita that he will give one more stone, when he will get. This last conversation that number represented by -1. So it is possible to find infinite number of numbers as natural number. The collection of all number which discussued above is called Integer and detoted as I or Z.

The mathematicaly Natural number, Whole number, Integar denoted and written as:

- $N = \{1, 2, 3, 4, \dots\}$

- $W = \{0, , 2, 3, 4, 5, \dots\}$

1. NUMBER SYSTEM

- $Z = \{\cdots -5, -4, -3, -2, -1, 0, 1, 2, 3, 4, 5, \ldots\}$

Properties of *Natural numbers*:

1. Sum of two *natular numbers* is always a *natular number*.

2. Product of two *natular numbers* is always a *natular number*.

3. Suntraction of two *natular numbers* need not be a *natular numbers*.

4. division of two *natular numbers* need not be a *natular number*.

Properties of *whole number*:

1. Sum of two *whole numbers* is always a *whole number*.

2. Product of two *whole numbers* is always a *whole number*.

3. Subtraction of two *whole numbers* need not be a *whole number*.

4. division of two *whole numbers* need not be a *whole number*.

Properties of *integer* :

1. Sum of two *integers* is always an *integer*

2. Product of two *integers* is always an *integer*

3. Suntraction of two *integers* always an *integer*

4. division of two *integers* need not be an *integer*

1. NUMBER SYSTEM

1.2 Rational numbers

We know that divison of any two integers need not be an integer, if we collect all integers and all numbers which is generated with division by two integers is called rational number and denoted by Q. Mathematicaly it is defined as,

Definition 1.2.1. A number r is called a **rational number**, if it can be written in the form $\frac{p}{q}$, where p and q are integer and $q \neq 0$.

Example 1.2.2.   i $1, 2, 4, -1, -2, -3$

ii $\frac{3}{4}, \frac{8}{7}$

iii $.2, .257, .24578950,$

iv $.\overline{5}, .\overline{3578},$

Equivalent rational number: The representation of a rational number is not unique, so the numbers that represent a rational number is called **equivalent rational numbers** or fractions.

Example 1.2.3.   i $\frac{3}{5} = \frac{9}{15} = \frac{12}{20}$

ii $\frac{5}{4} = \frac{10}{8}$

1. NUMBER SYSTEM

Remark: Every rational number has infinite number of equivalent rational number. Then all equivalent rational numbers represent a rational number $\frac{p}{q}$, where p and q has co-prime or theire gcd is 1.

Example 1.2.4. Are the following example true or false ? Give reason for your answers.

i Every whole number is a natural number.

ii Every integer is a rational number.

iii Every rational number is an integer.

iv Every rational number is a natural number.

v Every natural number is an integer.

vi Every integer is a natural number.

Solution:

i False, because zero is whole number, but not a natural number.

ii True, because every integer n can be expressed in the form $\frac{n}{1}$, which is a rational number.

iii False, at least one number $\frac{1}{2}$ is a rational but not integer.

1. NUMBER SYSTEM

iv False, because there is at least one number $\frac{1}{2}$ which is rational number, but not natural number.

v True, because every natural number is a positive integer.

vi False, because there is an integer -1, which is not a natural number.

Method to find a rational number between two rational numbers:

(1). Frist mathod to find a rartional number between two rational number.

Example 1.2.5. Find the five rational numbers between 1 and 2.

Solution :

We know that to find a rational number between r and s, you can add r and s and divide the sum by 2, that is $\frac{r+s}{2}$ lise between r and s. So, $\frac{3}{2}$ is a number between 1 and 2. Contune in this manner to find next four rational numbers between 1 and 2. These foue numbers $\frac{5}{4}, \frac{11}{8}, \frac{13}{8}, \frac{7}{4}$.

(2) Second method or One step method to find rational numbers betweet two rational number.

Solution : We want to five numbers between 1 and 2, we write 1 and 2 as equivalent rational with demoninator $5+1=6$, i.e., $1=\frac{6}{6}$ and $2=\frac{12}{6}$. Then write the numbers which numinatar are $6+1$, adding one by one upto 12 and denominator fixed 6. Get $\frac{7}{6}, \frac{8}{6}, \frac{9}{6}, \frac{10}{6}, \frac{11}{5}$ five rational numbers between 1 and 2.

1. NUMBER SYSTEM

Let a and b two rational number, then N rational number, say r_n between a and b given as:

$$r_n = \frac{a(N+1) + n(b-a)}{N+1}$$

Where $n = 1, 2, 3, \ldots N$.

1.3 Solutin of Exercise 1.1

Problem 1.3.1. Is zero a rational number ? Can you write it in the form $\frac{p}{q}$, where p and q are integer and $q \neq 0$.

Solution: Yes, because 0 can be written as $\frac{0}{q}, q \neq 0$.

Problem 1.3.2. Find six rational numbers between 3 and 4.

Solution: We know that the formula to find N rational number between a and b is

$$r_n = \frac{a(N+1) + n(b-a)}{N+1}$$

Where $n = 1, 2, 3, \ldots N$. So,

$$r_1 = \frac{3(6+1) + 1 \cdot 1}{6+1} = \frac{3 \cdot 7 + 1}{7} = \frac{22}{7}$$

Similarly

$$r_2 = \frac{23}{7}$$

$$r_3 = \frac{24}{7}$$

1. NUMBER SYSTEM

$$r_4 = \frac{25}{7}$$

$$r_5 = \frac{25}{7}$$

$$r_6 = \frac{26}{7}.$$

Therefore six rartional number between 3 and 4 are $\frac{22}{7}, \frac{23}{7}, \frac{24}{7}, \frac{25}{7}, \frac{26}{7}, \frac{27}{7}$.

Problem 1.3.3. Find five rational number between $\frac{3}{5}$ and $\frac{4}{5}$.

Solution: From second method, we have

$$r_n = \frac{a(N+1) + n(b-a)}{N+1}$$

Where $n = 1, 2, 3, \ldots N$.

$$r_1 = \frac{\frac{3}{5}(5+1) + 1\left(\frac{4}{5} - \frac{3}{5}\right)}{5+1} = \frac{\frac{3 \times 6}{5} + 1\frac{1}{5}}{6} = \frac{19}{30}$$

Similarly

$$r_2 = \frac{20}{30}$$

$$r_3 = \frac{21}{30}$$

$$r_n = \frac{22}{30}$$

$$r_n = \frac{23}{30}$$

The desire rational numbers are $r_1 = \frac{19}{30}, r_2 = \frac{2}{3}, r_3 = \frac{7}{10}, r_n = \frac{11}{15}, r_n = \frac{23}{30}$.

Problem 1.3.4. State whether the following statements are true or false. Give reasons for your answer.

1. NUMBER SYSTEM

    i . Every whole is a natural number.

    ii . Every integer is a whole number.

    iii . Every rational number is a whole number,

Solution:

    i . No, because 0 is not natural number but it belong to whole number.

    ii . No, because negative integer are not included whole number.

    iii . No, atleast one rational number as .2 is not whole number.

1.4 Irrational Numbers

Definition 1.4.1. A number 's' is called *irrational*, if it cannot be written in the form $\frac{p}{q}$, where p and q are integers and $q \neq 0$.

    Some examples: $\sqrt{2}, \sqrt{3}, \sqrt{17}, \pi, 0.10110111011110\ldots$.

Every real number is represented by a unique point on the number line. Also, every point on the number line represents a unique real number.

Example 1.4.2. Locate $\sqrt{2}$ on the number line.

1. NUMBER SYSTEM

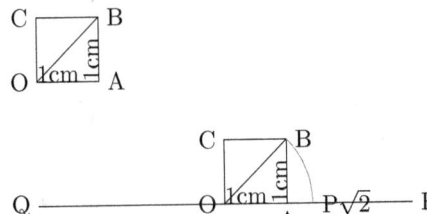

Solution: Consider a square OABC, with each side 1 unite length.

Consider a square OABC, with side 1 unit in length. Then according Pythagoras theorem the diagonal of square will be $\sqrt{2}$. Now identify the point whose distance equal to length OB, this point represent $\sqrt{2}$ on number line.

Example 1.4.3. Locate $\sqrt{3}$ on the number line.

Solution:

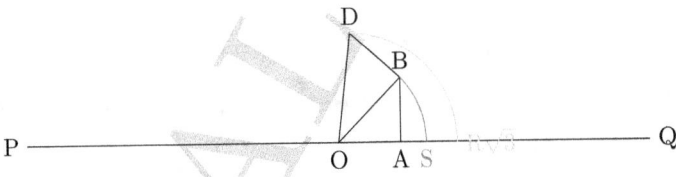

Construct AB of unit lenght perpendicular on line OA, which is also unit length. Therefore by Pythagoras theorem, we have $OB = \sqrt{1^2 + 1^2} = \sqrt{2}$. Similarally $OD = \sqrt{\sqrt{2}^2 + 1^2} = \sqrt{3}$. Now draw an arc with radius $\sqrt{3}$ cm at center 'O' which cut line segment PQ at R that represent $\sqrt{3}$ on number line.

1. NUMBER SYSTEM

1.5 Solution of Exercise 1.2

Problem 1.5.1. State whether the following statements are true or false. Justify your answers.

1. Every irrational number is a real number.

2. Every point on the number line is of the form $\sqrt{m}$, where m is a natural number.

3. Every real number is a rational number.

Solution:

1. Yes, real number is collection of all rational and irrartional numbers.

2. No, -1 is not square root of any natural number.

3. No, real number contains rational and irrational numbers.

Problem 1.5.2. Are the square roots of all positive integers irrational ? If not, give an example of the square root of a number that is a rational number.

Solution: No, $\sqrt{4}$ is 2, which is a rational number.

Problem 1.5.3. Show how $\sqrt{5}$ can be represented on the number line.

Solution: Frist draw a right angle triangle OAP_1 with OA and AP_1 are

1. NUMBER SYSTEM

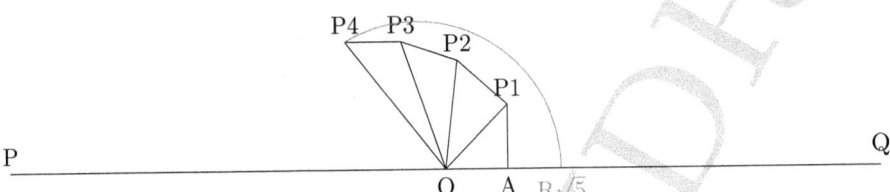

unit length, then by Pythagoras theorem $OP_1 = \sqrt{2}$. Similar construction draw perpendicular line at hypotenus of each triangle of unit length, the line segment $OP_4 = \sqrt{5}$. Next draw an arc of radius OP_4 which meet at point R on PQ. This is require point that represent $\sqrt{5}$ on number line.

1.6 Real Number and Decimal Expantions

We are familiar with decimal number system, each number can be represent in decimal number, So first observe some rational numbers and their decimal represntation.

Example 1.6.1. Find the decimal expansions of

  i $3 \overline{)10}$

  ii $7 \overline{)8}$

  iii $7 \overline{)1}$.

1. NUMBER SYSTEM

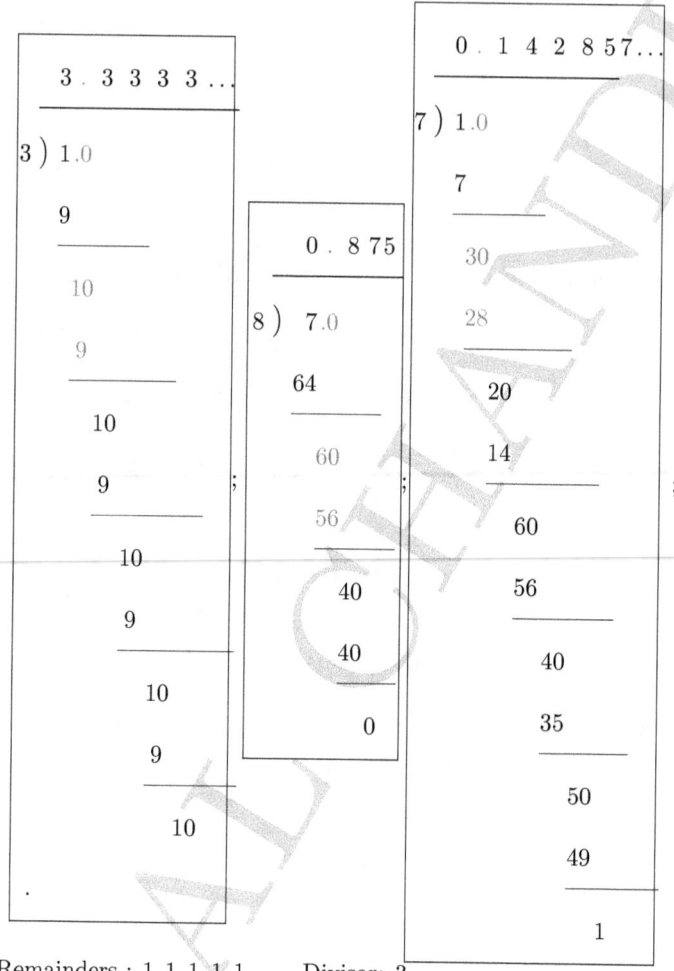

Remainders : $1,1,1,1,1,\ldots$, Divisor: 3,

Remainders : $6,4,0,$, Divisor: 8,

Remainders : $3,2,6,4,5,1,3,2,6,4,5,1,\ldots$, Divisor: 7.

Observation: If you closely analysize the example (1.6.1), you get follow-

1. NUMBER SYSTEM

ing facts;

1. The remainder either becomes 0 after certain a stage, or start repeating themseves.

2. The number of entries in the repeating string of remainders is less than the divisior.

3. If the remainders repeat, then we get a repeating block of digits in the quotient.

The above fact true for all rational number which in the form $\frac{p}{q}, q \neq 0$. So on division of p by q two cases arise either remainder zero or never become zero.

The remainder becomes zero: When remainder becomes zero after a certain stage, then in the decimal representations of rational number the decimal number terminates or ends after a finite number of steps. This decimal representation is called **terminating decimal**.

The remainder never becomes zero: When remainder never becomes zero, then it repeats after a certain stage, so we find a repeating block of digits in quotion. This representation of decimal called **non terminating recurring decimal**.

Example 1.6.2. Show that 3.142678 is a rational number.

1. NUMBER SYSTEM

Solution: We have $3.142678 = \frac{3.142678}{1000000}$, hence it is a rational number.

Example 1.6.3. Show that $0.333\cdots = 0.\overline{3}$, can expressed in the form $\frac{p}{q}$ where p and q are integer and $q \neq 0$.

Solution: Let x is $0.\overline{3}$. So,

$$x = 0.333\ldots$$

Now

$$10x = 10 \times (0.333\ldots) = 3.3333\ldots$$

next

$$10x = 3 + x$$

since $x = 0.333\ldots$,

Solving for x, we get

$$9x = 3, x = \frac{1}{3}$$

Example 1.6.4. Show that $1.27272727\cdots = 1.\overline{27}$ can be expressed in the form $\frac{p}{q}$, where p and q are integers and $q \neq 0$.

Solution: Let $x = 1.\overline{27}$, here two digit are repeating, so we multiply x by 100 to get,

$$100x = 1.272727\ldots$$

1. NUMBER SYSTEM

$$100x = 126 + 1.272727\cdots = 126 + x$$

So

$$100x - x = 126 \implies 99x = 126$$

$$x = \frac{126}{99} = \frac{14}{11}$$

Example 1.6.5. Show that $0.235353535\cdots = 0.2\overline{35}$ can be expressed in the form $\frac{p}{q}$, where p and q are integers and $q \neq 0$

Solution: Let $x = 0.2\overline{35}$, here 2 is not repeated but the block 35 repeating, so we multiply x by 100 to get,

$$100x = 23.53535\ldots$$

$$100x = 23.3 + 0.23535\cdots = 23.3 + x$$

So

$$100x - x = 23.3 \implies 99x = 23.3$$

$$x = \frac{233}{990}$$

$$x = \frac{233}{990} = 0.2\overline{35}$$

1. NUMBER SYSTEM

Result: *The decimal expansion of rational number is either terminating or non-terminating recurring. Moreover, a number whose decimal expantion is terminating or non-terminating recurring is rational.*

From above discussion, we recognise the decimal expantion of rational number, but except rational number the decimal representation is **non-terminating non-recurring**.

Result-2 : *The decimal expantion of an irrational number is non-terminating non recurring. Moreover, a number whose decimal expantion is non-terminating non-recurring is irrational.*

Example 1.6.6. Find an irrational number between $\frac{1}{7}$ and $\frac{2}{7}$.

Solution : We know that $\frac{1}{7} = 0.\overline{142857}$. So we can find observing the remainder of $\frac{1}{7} = 0$, we find $\frac{2}{7} = 0.\overline{285714}$. To find an irrational number between $\frac{1}{7}$ and $\frac{2}{7}$, means a number which is non-terminating non-recurring lying between them. One of them is $0.160160016000160000160000\ldots$.

1.7 Solution of Exercise 1.3

Problem 1.7.1. Write the following in decimal form and say what kind of decimal expantion each has :

i $\frac{36}{100}$,

1. NUMBER SYSTEM

ii $\dfrac{1}{11}$,

iii $4\dfrac{1}{8}$,

iv $\dfrac{3}{13}$,

v $\dfrac{2}{11}$,

vi $\dfrac{329}{400}$.

Solution :

i
```
        0.36
100 ) 360
       300
       ---
       600
       600
       ---
         0
```

ii
```
         0.999 ...
      ----------
   11 ) 100
         99
         ---
         100
          99
         ---
         100
          99
         ---
           1
```

The $\dfrac{36}{100} = .36$, represents terminating decimal next the fraction $\dfrac{1}{11} = .9999\ldots$, represents non terminating recurring decimal.

1. NUMBER SYSTEM

$$\begin{array}{r} 4.125 \\ 8\overline{)33} \\ 32 \\ \hline 10 \\ 8 \\ \hline 20 \\ 16 \\ \hline 40 \\ 40 \\ \hline 0 \end{array}$$

iii

$$\begin{array}{r} 0.2376153846\ldots \\ 13\overline{)30} \\ 26 \\ \hline 40 \\ 39 \\ \hline 100 \\ 91 \\ \hline 90 \\ 78 \\ \hline 20 \\ 13 \\ \hline 70 \\ 65 \\ \hline 50 \\ 39 \\ \hline 110 \\ 104 \\ \hline 60 \\ 52 \\ \hline 80 \\ 78 \\ \hline 2 \end{array}$$

iv

1. NUMBER SYSTEM

The fraction $4\frac{1}{8} = 4.125$ is representations of terminating decimal. The fraction $\frac{3}{13} = 0.2376153$ is representations of non-terminating recurring decimal.

```
              0.1818...
           _____
      11 ) 20                           0.825
                                     _____
            11                   400 ) 3290
           ____
            90                         3200
                                       ____
            88                          900
v                         vi
           ____                         800
            20                         ____
                                       2000
            11
                                       2000
           ____                        ____
             90                           0
             88
            ____
              2
```

The fraction $\frac{2}{11} = 0.1818\ldots$ is representations of non-terminating recurring decimal. The fraction $\frac{429}{400} = 4.125$ is representations of terminating decimal.

Problem 1.7.2. You know that $\frac{1}{7} = 0.\overline{142857}$. Can you predict what the decimal expansions of $\frac{2}{7}, \frac{3}{7}, \frac{4}{7}, \frac{5}{7}, \frac{6}{7}$ are, without actually doing the long division ? If so, how ?

1. NUMBER SYSTEM

Solution : After division of $\frac{1}{7} = 0.142857\ldots$ if you carefully studies the sequence of remainder and quotiont are corelated eachother as $1, 3, 2, 6, 4, 5$, and $1, 4, 2, 8, 5, 7$ repeated infinitaly times. So, the quotiont of fraction $\frac{2}{7}$ has contains in this ordered $.2, 8, 5, 7, 1, 4$, because the remainders has been apears in this formate also. Hence $\frac{2}{7} = 0.857142\ldots$. Simillarly $\frac{3}{7} = 0.428571\ldots, \frac{4}{7} = 0.571428\ldots, \frac{5}{7} = 0.714285\ldots, \frac{6}{7} = 0.857142$ will be,

Problem 1.7.3. Express the following in the form $\frac{p}{q}$, where p and q are integer and $q \neq 0$.

    i $0.\overline{6}$

    ii $0.4\overline{7}$

    iii $0.\overline{001}$

    Solution :

i   Let $x = 0.\overline{6}$. Since one digit is repeating, we multiply x by 10 to get

$$10x = 6.666666\ldots$$

$$10x = 6 + 0.6666666\ldots$$

$$10x = 6 + x$$

$$9x = 6,$$

1. NUMBER SYSTEM

which gives
$$x = \frac{2}{3}$$

ii Let $x = 0.4\overline{7}$. Since one digit is repeating, we multiply x by 10 to get

$$10x = 4.77777\ldots$$

$$10x = 4 + 0.3 + 0.4777777\ldots$$

$$10x = 4.3 + x$$

$$9x = 4.3,$$

which gives
$$x = \frac{43}{90}$$

iii Let $x = 0.\overline{001}$. Since three digit is repeating, we multiply x by 1000 to get

$$1000x = 1.001001001\ldots$$

$$1000x = 1 + 0.001001001\ldots$$

$$1000x = 1 + x$$

$$999x = 1,$$

which gives
$$x = \frac{1}{999}$$

1. NUMBER SYSTEM

Problem 1.7.4. Express 0.99999... in the form $\frac{p}{q}$, Are you surprised by your answer ? With your teacher and classmets discuss why ? the answer makes sense.

Solution :

Let $x = 0.\overline{9} = 0.999999\ldots$. Since one digit is repeating, we multiply x by 10 to get

$$10x = 9.999999\ldots$$

$$10x = 9 + 0.999999\ldots$$

$$10x = 9 + x$$

$$9x = 9,$$

which gives

$$x = \frac{9}{9} = 1$$

Problem 1.7.5. What can be maximum number of digits be in the repeating block of digits in the decimal expansion of $\frac{1}{17}$? Perform the division to check your answer.

Solution : The maximunm number of digits in repeating block of decimal expantion of $\frac{1}{17}$ can be 16.

1. NUMBER SYSTEM

```
            0.0588235294117647
      17 ) 100
            85
           ---
           150
           136
           ---
            140
            136
            ---
              40
              34
              ---
               60
               51
               ---
                90
                85
                ---
                 50
                 34
                 ---
                 160
                 153
                 ---
                   70
                   68
                   ---
                    20
                    17
                    ---
                     30
                     17
                     ---
                     130
                     119
                     ---
                      110
                      102
                      ---
                       80
                       68
                       ---
                       120
                       119
                       ---
                         1
```

1. NUMBER SYSTEM   24

Problem 1.7.6. Write three numbers whose decimal expansions are non-terminating non-recurring.

Solution : The following decimal numbers are non-terminating non-recurring.

i $0.0100100010001000001...$

ii $0.18018001800018000018...$

iii $0.27027002700027000027...$

Problem 1.7.7. Find three different irrational numbers between the rational numbers $\frac{5}{7}$ and $\frac{9}{11}$.

Solution : We have two given numbers as $\frac{5}{7} = 0.\overline{71428}$ and $\frac{9}{11} = 0.\overline{81}$, then non-terminating and non-recurring decimal number between these two numbers give irrartional numbers that lise bwtwen them also. As

i $0.71071007100071000071...$

ii $0.72072007200072000072...$

iii $0.73073007300073000073...$

Problem 1.7.8. Classify the following numbers as rational or irrational:

i $\sqrt{23}$

1. NUMBER SYSTEM

    ii $\sqrt{225}$

    iii 0.3796

    iv 7.478478...

    v 1.101001000100001...

Solution: In the given number following are rational numbers 0.3796, 7.478478... and irrational numbers are $\sqrt{225}$, $\sqrt{23}$, 1.101001000100001....

1.8 Operations and Real Numbers

You know that the rational numbers satisfy the `commutative`, `associative` and `distributive law for addition and multiplication`. Moreover, if we add, subtract, multiply or divide (except zero) two rational numbers, we still get a rational number. This property is called that rational numbers are **'closed'** with respect to addition, subtraction, multiplication and division. However the sum, difference and product of irrational numbers are need not irrartional numbers. We discuss many example to clearify this notion.

Example 1.8.1. Check whether $(\sqrt{6}) + (-\sqrt{6})$, $(\sqrt{3}) \cdot (\sqrt{3})$, and $\dfrac{\sqrt{17}}{\sqrt{17}}$ are rational or irrational numbers.

1. NUMBER SYSTEM

Solution:

$$\left(\sqrt{6}\right) + \left(-\sqrt{6}\right) = \sqrt{6} - \sqrt{6} = 0,$$

$$\left(\sqrt{3}\right) \cdot \left(\sqrt{3}\right) = \sqrt{3}.\sqrt{3} = 3,$$

$$\frac{\sqrt{17}}{\sqrt{17}} = 1.$$

All these are integer. So, all these are rational numbers.

Example 1.8.2. Check whether $7\sqrt{5}, \dfrac{7}{\sqrt{5}}, \sqrt{2}+21, \pi-2$ are irrational or not.

Solution: We know that $\sqrt{5} = 2.236\ldots, \sqrt{2} = 1.1412\ldots, \pi = 3.1415\ldots$

Then $7\sqrt{5} = 15.652\ldots, \dfrac{7}{\sqrt{5}} = \dfrac{7\sqrt{5}}{\sqrt{5}\sqrt{5}} = \dfrac{7\sqrt{5}}{5} = 3.1304\ldots$

$\sqrt{2}+21 = 22.4142\ldots, \pi-2 = 1.1415\ldots$ All these are non-terminating non-recurring decimals. So, all these are irrational numbers.

Result:

i The sum, difference and product of irrational numbers are need not be an irrartional number.

ii The *nth* root of an positive irrartional number need not be an irrational number.

Example 1.8.3. Add $2\sqrt{2} + 5\sqrt{5}$ and $\sqrt{2} - 3\sqrt{3}$.

Solution:

$$\left(2\sqrt{2} + 5\sqrt{5}\right) + \left(\sqrt{2} - 3\sqrt{3}\right) = \left(2\sqrt{2} + \sqrt{2}\right) + \left(5\sqrt{3} - 3\sqrt{3}\right)$$

1. NUMBER SYSTEM

$$= (2+1)\sqrt{2} + (5-3)\sqrt{3} = 3\sqrt{2} + 2\sqrt{3}$$

Example 1.8.4. Multiply $6\sqrt{5}$ by $2\sqrt{5}$.

Solution: $6\sqrt{5} \times 2\sqrt{5} = 6 \times 2 \times \sqrt{5} \times \sqrt{5} = 12 \times 5 = 60$

Example 1.8.5. Divide $8\sqrt{15}$ by $2\sqrt{3}$.

Solution: $\dfrac{8\sqrt{15}}{2\sqrt{3}} = \dfrac{8\sqrt{3}\sqrt{5}}{2\sqrt{3}} = 4\sqrt{5}$

From these example we can conclude the followinf fact

i The sum or difference of a rational number and irrational numbers is an irrational number.

ii The product or quotient of a non zero rational number with an irrational number is an irrational number.

iii If we add, subtract, multiply or divide two irrational numbers the result may be rational or irrational number.

Squra root of positive rational number

Let $a > 0$ be a real number. Then $\sqrt{a} = b$ means $b^2 = a$ and $b > 0$. Now we will show, how to find $\sqrt{x}$ for any given positive real number x geometricaly. More generaly, to find $\sqrt{x}$ for any positive real number x, we mark B so that

1. NUMBER SYSTEM

$AB = x$ units, and as in Fig.(1.8.1) mark C so that $BC = 1$ unit. Then we find $BD = \sqrt{x}$. We can prove this result using the pythagoras Theorem.

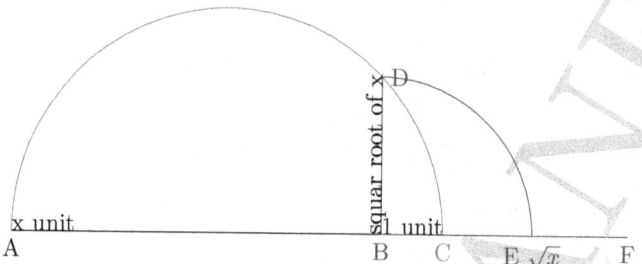

Figure 1.8.1: Representation of $\sqrt{x}$

Identites related to squar root

Let a and b be positive real numbers. Then

i $\sqrt{ab} = \sqrt{a}\sqrt{b}$

ii $\sqrt{\dfrac{a}{b}} = \dfrac{\sqrt{a}}{\sqrt{b}}$

iii $\left(\sqrt{a}+\sqrt{b}\right)\left(\sqrt{a}-\sqrt{b}\right) = a-b$

iv $\left(a+\sqrt{b}\right)\left(a-\sqrt{b}\right) = a^2 - b$

v $\left(\sqrt{a}+\sqrt{b}\right)\left(\sqrt{c}+\sqrt{d}\right) = \sqrt{ac}+\sqrt{ad}+\sqrt{bc}+\sqrt{bd}$

vi $\left(\sqrt{a}+\sqrt{b}\right)^2 = a + 2\sqrt{ab} + b$

Example 1.8.6. Simplfy the following expressions:

1. NUMBER SYSTEM

i $(5+\sqrt{7})(2+\sqrt{5})$

ii $(5+\sqrt{5})(5-\sqrt{5})$

iii $(\sqrt{3}+\sqrt{7})^2$

iv $(\sqrt{11}-\sqrt{7})(\sqrt{11}-\sqrt{7})$

Solution:

i $(5+\sqrt{7})(2+\sqrt{5}) = 10+5\sqrt{5}+2\sqrt{7}+\sqrt{35}$

ii $(5+\sqrt{5})(5-\sqrt{5}) = 5^2 - (\sqrt{5})^2 = 25 - 5 = 20$

iii $(\sqrt{3}+\sqrt{7})^2 = (\sqrt{3})^2 + 2\sqrt{3}\sqrt{7} + (\sqrt{7})^2 = 3 + 2\sqrt{21} + 7 = 10 + 2\sqrt{21}$

iv $(\sqrt{11}-\sqrt{7})(\sqrt{11}-\sqrt{7}) = (\sqrt{11})^2 - (\sqrt{7})^2 = 11 - 7 = 4$

Remark: Note that after 'simplification' the above example, we get that the expression should be written as the sum of a rational and an irrational number. Lte us see, if we can 'rationalise the denomenator of a fraction, that maens to make the denominator into a rational number.

Example 1.8.7. Rationalise the denominator of $\dfrac{1}{\sqrt{2}}$.

Solution: We want to write $\dfrac{1}{\sqrt{2}}$ as an equivalent fraction in which the denominator is rational number. We know that $\sqrt{2}.\sqrt{2}$ is a rational number. Also we know that multiplication non zero number in numenator and denominator gives an equivalent fraction. So, cosidaring these fact we get,

$$\frac{1}{\sqrt{2}} = \frac{1}{\sqrt{2}} \times \frac{\sqrt{2}}{\sqrt{2}} = \frac{\sqrt{2}}{2}$$

Now it is easy to locate $\frac{1}{\sqrt{2}}$ on number line.

Example 1.8.8. Rationalise the denominator of $\frac{1}{2-\sqrt{3}}$.

Solution: To rationalise the $\frac{1}{2-\sqrt{3}}$, multiply and divide by $2-\sqrt{3}$ to get $\frac{1}{2+\sqrt{3}} \times \frac{2-\sqrt{3}}{2-\sqrt{3}} = \frac{2-\sqrt{3}}{4-3} = 2-\sqrt{3}$.

Example 1.8.9. Rationalise the denominator of $\frac{5}{\sqrt{3}-\sqrt{5}}$.

Solution: Here

$$\frac{5}{\sqrt{3}-\sqrt{5}} = \frac{5}{\sqrt{3}-\sqrt{5}} \times \frac{\sqrt{3}+\sqrt{5}}{\sqrt{3}+\sqrt{5}} = \frac{5(\sqrt{3}+\sqrt{5})}{3-5} = \frac{-5}{2}(\sqrt{3}+\sqrt{5}).$$

Example 1.8.10. Rationalise the denominator of $\frac{1}{7+3\sqrt{2}}$.

Solution:

$$\frac{1}{7+3\sqrt{2}} = \frac{1}{7+3\sqrt{2}} \times \frac{7-3\sqrt{2}}{7-3\sqrt{2}} = \frac{7-3\sqrt{2}}{49-18} = \frac{7-3\sqrt{2}}{31}.$$

The process of converting a real number to an equivalent expression whose denominator is a rational number is called **rationalising the denominator**.

1. NUMBER SYSTEM

1.9 Solution of Exercise 1.4

Problem 1.9.1. Classsify the following numbers as rational or irrational:

i $2 - \sqrt{5}$

ii $(3 + \sqrt{23}) - \sqrt{23}$

iii $\dfrac{2\sqrt{7}}{7\sqrt{7}}$

iv $\dfrac{1}{\sqrt{2}}$

v 2π

Solution:

i Irrational number. The decimal representation of $2 - \sqrt{5}$ is non terminating non-recurring decimal.

ii Rational number. After simplify of $(3 + \sqrt{23}) - \sqrt{23} = 3$, which is a rational number.

iii Rational number. After simplify $\dfrac{2\sqrt{7}}{7\sqrt{7}} = \dfrac{2}{7}$, which is a rational number.

iv Irrational number. After simplify $\dfrac{1}{\sqrt{2}} = \dfrac{\sqrt{2}}{2}$, which gives non terminating non-recurring decimal reprcscntion.

v Irrational number. It also represents non terminating non-recurring decimal.

Problem 1.9.2. Simplify each of the following expression:

i
$$\left(3+\sqrt{3}\right)\left(2+\sqrt{2}\right)$$

ii
$$\left(3+\sqrt{3}\right)\left(3-\sqrt{3}\right)$$

iii
$$\left(\sqrt{5}+\sqrt{2}\right)^2$$

iv
$$\left(\sqrt{5}-\sqrt{2}\right)\left(\sqrt{5}+\sqrt{2}\right)$$

Solution:

i $\left(3+\sqrt{3}\right)\left(2+\sqrt{2}\right) = 3\times 2 + 3\sqrt{2} + 2\sqrt{3} + \sqrt{2}\sqrt{3} = 6 + 3\sqrt{2} + 2\sqrt{3} + \sqrt{2}\sqrt{3}$

According identities (v) we simplify the problems.

ii Applying identies iv, $\left(3+\sqrt{3}\right)\left(3-\sqrt{3}\right) = 3^2 - \left(\sqrt{3}\right)^2 = 9 - 3 = 6$

iii Applying identies vi, $\left(\sqrt{5}+\sqrt{2}\right)^2 = \left(\sqrt{5}\right)^2 + 2\sqrt{5}\sqrt{2} + \left(\sqrt{2}\right)^2 = 5 + \sqrt{10} + 2 = 7 + \sqrt{10}$

iv Applying identies iii, $\left(\sqrt{5}-\sqrt{2}\right)\left(\sqrt{5}+\sqrt{2}\right) = \left(\sqrt{5}\right)^2 - \left(\sqrt{2}\right)^2 = 5 - 2 = 3$

1. NUMBER SYSTEM

Problem 1.9.3. Recall, π is defined as the ratio of the circumference (say c) of a circle to its diameter (say d). That is $\pi = \dfrac{c}{d}$. This seems to contradict the fact that π is irrational. How will you resolve this contradiction ?

Solution: The ratio of circumference (say c) and diameter (say d) equivalt to π as $\pi \sim \dfrac{c}{d}$.

Problem 1.9.4. Represent $\sqrt{9.3}$ on number line.

Solution: Draw the line $AB = 9.3$ cm add $BC = 1$ cm, then draw semicircle as diameter $AC = 10.3$ and perpendicular line BD where point D on semicircle. We get required $BD = \sqrt{9.3}$ cm.

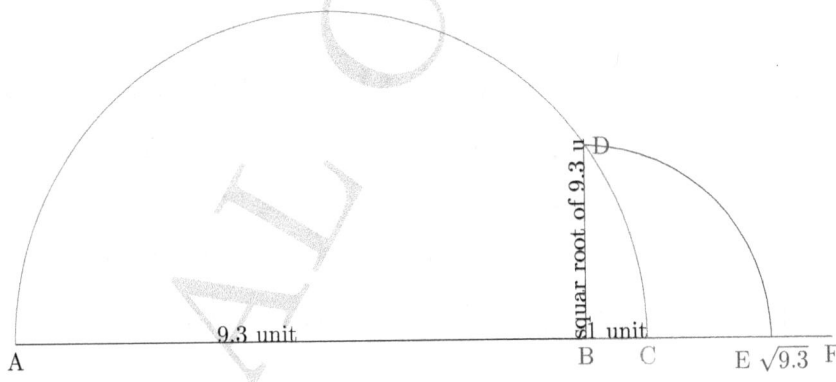

Problem 1.9.5. Rationalise the denominators of the following:

i $\dfrac{1}{\sqrt{7}}$

1. NUMBER SYSTEM

ii $\dfrac{1}{\sqrt{7}-\sqrt{6}}$

iii $\dfrac{1}{\sqrt{5}+\sqrt{2}}$

iv $\dfrac{1}{\sqrt{7}-2}$

Solution:

i $\dfrac{1}{\sqrt{7}} = \dfrac{1}{\sqrt{7}} \times \dfrac{\sqrt{7}}{\sqrt{7}} = \dfrac{\sqrt{7}}{7}$

ii $\dfrac{1}{\sqrt{7}-\sqrt{6}} = \dfrac{1}{\sqrt{7}-\sqrt{6}} \times \dfrac{\sqrt{7}+\sqrt{6}}{\sqrt{7}+\sqrt{6}} = \dfrac{\sqrt{7}+\sqrt{6}}{7-6} = \sqrt{7}+\sqrt{6}$

iii $\dfrac{1}{\sqrt{5}+\sqrt{2}} = \dfrac{1}{\sqrt{5}+\sqrt{2}} \times \dfrac{\sqrt{5}-\sqrt{2}}{\sqrt{5}-\sqrt{2}} = \dfrac{\sqrt{5}-\sqrt{2}}{5-2} = \dfrac{\sqrt{5}-\sqrt{2}}{3}$

iv $\dfrac{1}{\sqrt{7}-2} = \dfrac{1}{\sqrt{7}-2} \times \dfrac{\sqrt{7}+2}{\sqrt{7}+2} = \dfrac{\sqrt{7}+2}{7-4} = \dfrac{\sqrt{7}+2}{3}$

Remark: The conjugate of $\sqrt{a}+\sqrt{b}$ is $\sqrt{a}-\sqrt{b}$. To rationalise denominator, multiply conjugate of denominator in numenator and denominator then simplify it.

1.10 Law of Exponent for Real Numbers

In expression a^n, a is a rational number and n is a whole numbers. For the espression a^n, a called the base and n are the exponents. The exponent follow the following rules

i $a^m . a^n = a^{m+n}$

1. NUMBER SYSTEM

ii $(a^m)^n = a^{mn}$

iii $\dfrac{a^m}{a^n} = a^{m-n}, m > n$

iv $a^m b^m = (ab)^m$

v $\dfrac{1}{a^n} = a^{-n}$

vi $a^0 = 1$

Example 1.10.1. See the following

i $17^2 \cdot 17^{-5} = 17^{2-5} = 17^{-3} = \dfrac{1}{17^3}$

ii $\left(5^2\right)^{-7} = a^{2\times(-7)} = 5^{-14}$

iii $\dfrac{23^{-10}}{23^7} = 23^{-10-7} = 23^{-17}$,

iv $(7)^{-3} \cdot (9)^{-3} = (7.9)^{-3} = (63)^{-3}$

1.11 Law of Extended Exponents for Real Numbers

Now we define $\sqrt[n]{a}$ for a real number $a > 0$ as follow: Let $a > 0$ be a real number and n a positive integer. Then $\sqrt[n]{n} = b$, if $b^n = a$ and $b > 0$. In the language of exponents, we define $\sqrt[n]{a} = a^{\frac{1}{n}}$. So, in particuler $\sqrt[3]{5} = 3^{\frac{1}{5}}$.

1. NUMBER SYSTEM

Let $a > 0$ be a real number. Let m and n be integer such that m and n have no common factor other than 1 and $n > 0$. Then

$$a^{\frac{m}{n}} = \left(\sqrt[n]{a}\right)^n = \sqrt[n]{a^m}$$

Now we have following extended law of exponents:

Let $a > 0$ be a real number and p and q be rational numbers. Then, we have

i $a^p \cdot a^q = a^{p+q}$

ii $(a^p)^q = a^{pq}$

iii $\dfrac{a^p}{a^q} = a^{p-q}$

iv $a^p b^p = (ab)^p$

Example 1.11.1. Simplify

i $2^{\frac{2}{3}} \cdot 2^{\frac{2}{3}}$

ii $\left(3^{\frac{1}{5}}\right)^4$

iii $\dfrac{7^{\frac{1}{5}}}{7^{\frac{1}{3}}}$

iv $13^{\frac{1}{5}} \cdot 17^{\frac{1}{5}}$

Solution: Simplify

1. NUMBER SYSTEM

i By law i,
$$2^{\frac{2}{3}} \cdot 2^{\frac{2}{3}} = 2^{\frac{2}{3}+\frac{1}{3}} = 2^{\frac{3}{3}} = 2^1 = 2$$

ii By law ii
$$\left(3^{\frac{1}{5}}\right)^4 = 3^{\frac{4}{5}}$$

iii By law iii
$$\frac{7^{\frac{1}{5}}}{7^{\frac{1}{3}}} = 7^{\frac{1}{5}-\frac{1}{3}} = 7^{\frac{3-5}{15}} = 7^{\frac{-2}{15}}$$

iv By law of iv,
$$13^{\frac{1}{5}} \cdot 17^{\frac{1}{5}} = (13 \times 17)^{\frac{1}{5}} = 221^{\frac{1}{5}}$$

1.12 Solution of Exercise 1.5

Problem 1.12.1.   i $64^{\frac{1}{2}}$

ii $32^{\frac{1}{5}}$

iii $125^{\frac{1}{3}}$

Solution:

i $64^{\frac{1}{2}} = (8^2)^{\frac{1}{2}} = 8^{2 \times \frac{1}{2}} = 8^1 = 8$

ii $32^{\frac{1}{5}} = (2^5)^{\frac{1}{5}} = 5^{5 \times \frac{1}{5}} = 5^1 = 5$

1. NUMBER SYSTEM   38

iii $125^{\frac{1}{3}} = \left(5^3\right)^{\frac{1}{3}} = 5^{3\times\frac{1}{3}} = 5^1 = 5$

Problem 1.12.2. Find:

i $9^{\frac{3}{2}}$

ii $32^{\frac{2}{5}}$

iii $16^{\frac{3}{4}}$

iv $125^{-\frac{1}{3}}$

Solution:

i $9^{\frac{3}{2}} = \left(3^2\right)^{\frac{3}{2}} = 3^{2\times\frac{3}{2}} = 3^3 = 27$

ii $32^{\frac{2}{5}} = \left(2^5\right)^{\frac{2}{5}} = 2^{5\times\frac{2}{5}} = 5^2 = 25$

iii $16^{\frac{3}{4}} = \left(2^4\right)^{\frac{3}{4}} = 2^{4\times\frac{3}{4}} = 2^3 = 8$

iv $125^{-\frac{1}{3}} = \left(5^3\right)^{-\frac{1}{3}} = 5^{3\times\left(-\frac{1}{3}\right)} = 5^{-1} = \frac{1}{5}$

Problem 1.12.3. Simlify:

i $2^{\frac{2}{3}}.2^{\frac{1}{5}}$

ii $\left(\frac{1}{3^3}\right)^7$

1. NUMBER SYSTEM

iii $\dfrac{11^{\frac{1}{2}}}{11^{\frac{1}{4}}}$

iv $7^{\frac{1}{2}} \cdot 8^{\frac{1}{2}}$

Solution:

i $2^{\frac{2}{3}} \cdot 2^{\frac{1}{5}} = 2^{\frac{2}{3}+\frac{1}{5}} = 2^{\frac{10+3}{15}} = 2^{\frac{13}{15}}$

ii $\left(\dfrac{1}{3^3}\right)^7 = \left(3^{-3}\right)^7 = 3^{-3\times 7} = 3^{-21}$

iii $\dfrac{11^{\frac{1}{2}}}{11^{\frac{1}{4}}} = 11^{\frac{1}{2}-\frac{1}{4}} = 11^{\frac{2-1}{4}} = 11^{\frac{1}{4}}$

iv $7^{\frac{1}{2}} \cdot 8^{\frac{1}{2}} = (7 \times 8)^{\frac{1}{2}} = 56^{\frac{1}{2}}$

2 POLYNOMIALS

2.1 Introduction

In this chapter, we shall study with a particular type of algebraic expression, called *polynamial*, and the *terminology* related to it. We shall also study the *Remainder Theorem* and *Factor Theorem* and their use in the factorisation of polynomials. Next we shall study the algebraic identites and their use in factorisation and evaluating some given expressions.

2.2 Polynomials in One Variable

All the expression are of the form (a constant) × x . Now suppose we want to write an expression which is (a constant) × (a variable) and we do not know what the constant is. In such cases, we write the constant as a, b, c etc. So the expression will be ax, say. The values of the constants remain the same throughout a particular situation, that is the values of the constants do not change in a given problem, but the value of a variable can keep changing.
Now consider your roof shape is a squre and length of side is 5 unit. What is its perimeter ? Exactly it is $4 \times 5 = 20$ unit. Similarly the area of your roof is $5^2 = 25$ squar unit. If length side of your roof is variable x, then perimeter and area will be $4x$ unit and x^2 unit. Now we can write other many example

2. POLYNOMIALS

like this as $2x+5$, x^2+3x+1, x^3+x^2+9x+1, these algebraic expression has an intresting thing that the power of variable is non negative integer. The algebraic expression in which the power of variable is a non negative integer (whole number) is called *polynamials*. The polynomial which contains only one variable is called *polynomials of one variable*. The expression in a polynamial which is seperated by $+$ or $-$ are called a **terms** of polynomial.

In the polynomial x^3+x^2+9x+1, the x^3, x^2, and $9x$, 1 are the terms of polynamial. The each terms of a polynomial can be express as multiplication of a real number and variable with power (whole number), this real number is called **ceofficients** of variable with power (whole number). Here the coefficient of x^3, x^2, and x, x^0 are $1, 1, 9, 1$ in the polynomial x^3+x^2+9x+1. Now the 2 is also a polynomial and 2 is the ceofficient of x^0.

Example 2.2.1. Categaries the following algebraic expression in polynamial and non polynomial.

- $2+x$, this is polynomial.

- $\frac{1}{x}$, this is not polynomial because the power of variable negative integer -1.

- $\sqrt[3]{x}+x^2$ is not a polynomial because the power of variable $\frac{1}{3}$ which is not non negative integer.

2.3 Type of polynomial

Type of polynomial on the basis of term:

1. A polynamial in which no variable seems to apear is called **constant polynamial**, as $1, -1, 0$ etc.

2. A constant polynamial which is always 0 is called **zero polynomial**, as 0.

3. The polynomials having only one term are called **monomials**, as x, $2x$.

4. A polynomials having only two terms is called **binomial**, as $x+1$.

5. apolynomials having only three terms is called **trinomial**, as x^2+3x+5.

Remark:

1. A polynamial have only finite terms.

2. Each zero polynomial is a constant polynamail but every constant polynomialis not zero polynamial.

3. Each constant polynomial is monomial.

Degree of polynomial: The highest power of the variable present in a polynamial is called the *degree of the polynamial*.

2. POLYNOMIALS

Example 2.3.1. Find the degree of each polynomial

i $x^5 + x^4 + 3$

ii $2 - y^2 - y^3 + 2y^8$

iii 3

Solution:

i The highest power of the variable is 5. so the degree of the polynomial is 5.

ii The highest power of the variable is 8. so the degree of the polynomial is 8.

iii The term in the polynomial is only 3 which can be written as $3x^0$. So the exponent of x is 0. therefore, the degree of polynamial is 0.

Type of polynomial on the basis of degree:

1. A polynamial of dgree one is called **linear polynomial** as $x + 1$, x. The general form of a linear polynamials in one variable is $ax + b$, where $a \neq 0, b$ are real numbers.

2. POLYNOMIALS

2. A polynamial of degree two is called a **quadratic polynomial**, as $2x^2 + x + \pi$, $x^2 + 8$, x^2. In one variable the gerernal form of aqudratic polynomial is $ax^2 + bx + c$, where $a \neq 0$ and b, c are real numbers.

3. A polynomial of degree three is called **cubic polynomial**, as $\sqrt[3]{2}x^3 + x^2 + x + 2$, $x^3 + \pi x^2 + 3x + 9$, x^3, $x^3 + x^2$. In one variable the general form of cubic polynomial is $ax^3 + bx^2 + cx + d$, where $a \neq 0$ and b, c, d are real numbers.

Remark:

i The degree of zero polynomial is not define.

ii The linear polynomial having at most two terms.

iii The qudratic polynomial having at most three terms.

iv The cubic polynomial having at most four terms.

2.4 Solution of Exercise 2.1

Problem 2.4.1. Which of the following expressions are polynomials in one variable and which are not ? State reason for your answer.

i $4x^2 - 3x + 7$

2. POLYNOMIALS

ii $y^2 + \sqrt{2}$

iii $3\sqrt{t} + t\sqrt{2}$

iv $y + \frac{1}{y}$

v $x^{10} + y^3 + t^{50}$

Solution

i It is a polynomial in one variable. The exponent of variable is non negative integer.

ii It is a polynomial in one variable. The exponent of variable is non negative integer.

iii It is not a polynomial in one variable. The exponent of variable is fraction.

iv It is not a polynomial in one variable. The exponent of variable is negative integer (-1).

v It is not a polynomial in one variable. There are three variable present in polynomial.

Problem 2.4.2. Write the coefficient of x^2 in each of the following:

i $2 + x^2 + x$

2. POLYNOMIALS

    ii $\ 2 - x^2 + x^3$

    iii $\ \frac{\pi}{2}x^2 + x$

    iv $\ \sqrt{2}x - 1$

Solution:

    i The coefficient of x^2 is 1.

    ii The coefficient of x^2 is -1.

    iii The coefficient of x^2 is $\frac{\pi}{2}$.

    iv The coefficient of x^2 is 0.

Problem 2.4.3. Give one example each of a binomial of degree 35 and of a momomial of degree 100.

    Solution: The general form of binomials of degree 35 is $ax^{35} + b$, where $a, b \neq 0$ are real numbers. So polynomial is like $x^{35} + 1$. The general form of monomials of degree 100 is ax^{100}, where $a \neq 0$ is real numbers. So polynomial is like x^{100}.

Problem 2.4.4. Write the degree of each of the following polynomials:

2. POLYNOMIALS

Polynomial	Degree
(i) $5x^3 + 4x^2 + 7x$	3
(ii) $4 - y^2$	2
(iii) $5t - \sqrt{7}$	1
(iv) 3	0

Problem 2.4.5. Classify the following as linear, quadratic and cubic polynomials:

i $x^2 + x$

ii $x - x^3$

iii $y + y^2 + 4$

iv $1 + x$

v $3t$

vi r^2

vii $7x^3$

Solution:

i Linear, the polynomial (iv) and (v) are linear polynomials.

ii Quadratic, the polynomial $(i), (iii), (vi)$ are quadratic polynomials.

2. POLYNOMIALS

iii Cubic, the polynomial $(ii), (vii)$ are cubic polynomials.

2.5 Zeros of a Polynomial

Consider the polynomial $p(x) = 2x^2 - 3x + 1$. If we put $x = 2$ in $p(x)$, we get

$$p(1) = 2 \times 2^2 - 3 \times 2 + 1$$

$$= 2 \times 4 - 6 + 1$$

$$= 8 - 5$$

$$= 3$$

This number 3 is called value of polynomial $p(x)$ at $x = 2$.

Values of a polynomial

Let $p(x)$ is polynomial in one variable x, then $p(k)$ is called *value* of polynomial at $x = k$, where k is any rael number. Now for which $x = k$ the value of polynomial $p(k)$ is zero, then $x = k$ is called **zero of polynomial**.

Remark:

i Zero of polynomial may not exist for each polynomial.

ii Zero of polynomial is not unique.

2. POLYNOMIALS

Example 2.5.1. Check whether -2 and $-1, 2$ are zeroes of the polynomial $x + 2$.

Solution: Let $p(x) = x+2$. Then $p(-2) = -2+2 = 0$, $p(-1) = -1+2 = 1$, $p(2) = 2+2 = 4$, Therefore, -2 is a zero of the polynomial $x + 2$, but $-1, 2$ are not.

Example 2.5.2. Find the zero of polynomial $p(x) = 3x + 1$.

Solution: Now we equate the polynomial $p(x)$ is zero and find out the value x. $p(x) = 0$, $3x + 1 = 0$, $3x = -1$, $x = \frac{1}{3}$. So, the zero of polynomial is $-\frac{1}{3}$.

Example 2.5.3. Verify whether 3 and 0 are zeroes of the polynomial $x^2 - 3x$.

Solution: We have polynomial $p(x) = x^2 - 3x$.

Then

$$p(3) = 3^2 - 3 = 9 - 9 = 0$$

$$p(0) = 0^2 - 3.0 = 0 - 0 = 0$$

Hence 3 and 0 bothe are zeroes of the polynomial $x^2 - 3x$.

2.6 Solution of Exercise 2.2

Problem 2.6.1. Find the value of the polynomial $5x - 4x^2 + 3$ at

i $x = 0$,

2. POLYNOMIALS

ii $x = -1$,

iii $x = 2$

Solution: Let $p(x) = 5x - 4x^2 + 3$

i $p(0) = 5.0 - 40^2 + 3 = 0 + 0 + 3 = 3$

ii $p(-1) = 5 \times (-1) - 4(-1)^2 + 3 = -5 - 4 + 3 = -9 + 3 = -6$

iii $p(2) = 5.2 - 4.2^2 + 3 = 10 - 4.4 + 3 = 10 - 8 + 3 = 5$

Problem 2.6.2. Find the value $p(0), p(1)$ and $p(2)$ for each of the following polynomial:

i $p(y) = y^2 - y = 1$

ii $p(t) = 2 + t + 2t^2 - t^3$

iii $p(x) = x^3$

iv $p(x) = (x - 1)(x + 1)$

Solution:

i Let $p(y) = y^2 - y + 1$,

$$p(0) = 0^2 - 0 + 1 = 1$$

$$p(1) = 1^2 - 1 + 1 = 1 + 0 = 1$$

2. POLYNOMIALS

$$p(2) = 2^2 - 2 + 1 = 4 - 2 + 1 = 3$$

ii Let $p(t) = 2 + t + 2t^2 - t^3$

$$p(0) = 2 + 0 + 2.0^2 - 0^3 = 2 + 0 + 0 - 0 = 2$$

$$p(1) = 2 + 1 + 2.1^2 - 1^3 = 2 + 1 + 2 - 1 = 4$$

$$p(2) = 2 + 2 + 2.2^2 - 2^3 = 4 + 2.4 - 8 = 4 + 8 - 8 = 4$$

iii Let $p(x) = x^3$

$$p(0) = 0^3 = 0$$

$$p(1) = 1^3 = 1$$

$$p(2) = 2^3 = 8$$

iv Let $p(x) = (x-1)(x+1)$

$$p(0) = (0-1)(0+1) = (-1).1 = -1$$

$$p(1) = (1-1)(1+1) = 0.1 = 0$$

$$p(2) = (2-1)(2+1) = 1.3 = 3$$

Problem 2.0.3. Verify whether the following are zeroes of the polynomial, indicated againist them.

2. POLYNOMIALS

i $p(x) = 4x + 1, x = -\frac{1}{3}$. Next

$$p\left(-\frac{1}{3}\right) = 3\left(-\frac{1}{3}\right) + 1 = -1 + 1 = 0$$

hence $-\frac{1}{3}$ is zero of polynomial.

ii Let $p(x) = 5x - \pi, x = \frac{4}{5}$, next

$$p\left(\frac{4}{5}\right) = 5.\frac{4}{5} - \pi = 4 - \pi$$

hence $\frac{4}{5}$ is not zero of the polynomial.

iii Let $p(x) = x^2 - 1, x = -1, 1$.

Now,

$$p(1) = 1^2 - 1 = 1 - 1 = 0$$

$$p(-1) = (-1)^2 - 1 = 1 - 1 = 0$$

Hence $1, -1$ bothe are zeroes of polynomial.

iv Let $p(x) = (x+1)(x-2), x = -1, 2$ Next

$$p(-1) = (-1+1)(-1-2) = 0.(-3) = 0$$

$$p(2) = (2+1)(2-2) = 3.0 = 0$$

Hence $-1, 2$ bothe are zeroes of the polynomial.

2. POLYNOMIALS

v Let $p(x) = x^2, x = 0$ Next

$$p(0) = 0^2 = 0$$

Hence 0 is the zero of polynomial.

vi Let $p(x) = lx + m, x = -\frac{m}{l}$, Next

$$p\left(-\frac{m}{l}\right) = l\left(-\frac{m}{l}\right) + m = -m + m = 0$$

Hence $-\frac{m}{l}$ is the zero of the polynomial.

vii Let $p(x) = 3x^2 - 1, x = -\frac{1}{\sqrt{3}}, \frac{2}{\sqrt{3}}$ Next

$$p\left(-\frac{1}{\sqrt{3}}\right) = 3\left(-\frac{1}{\sqrt{3}}\right)^2 - 1 = 3.\frac{1}{3} - 1 = 1 - 1 = 0$$

$$p\left(\frac{1}{\sqrt{3}}\right) = 3\left(\frac{1}{\sqrt{3}}\right)^2 - 1 = 3\frac{1}{3} - 1 = 1 - 1 = 0$$

Hence $-\frac{1}{\sqrt{3}}, \frac{1}{\sqrt{3}}$ are the zeroes of polynomial.

viii Let $p(x) = 2x + 1, x = \frac{1}{2}$. Next

$$p\left(\frac{1}{2}\right) = 2\frac{1}{2} + = 1 + 1 = 2$$

Hence $\frac{1}{2}$ is not zero of polynomial.

Problem 2.6.4. Find the zero of the polynomial in each of the following cases:

2. POLYNOMIALS

i Let $p(x) = x + 5$, consider k is the zero of the polynomial $p(x)$, then $p(k) = 0$.

$$p(k) = 0$$
$$k + 5 = 0 \implies k = -5$$

Hence zero of the $p(x)$ is -5.

ii Simirally (i), let $p(x) = x - 5$

$$p(k) = 0 \implies k - 5 = 0 \implies k = 5$$

So zero of $p(x)$ is 5.

iii Simirally (i), let $p(x) = 2x + 5$

$$p(k) = 0 \implies 2k + 5 = 0 \implies 2k = -5 \implies k = \frac{-5}{2}$$

. So zero of $p(x)$ is $-\frac{5}{2}$

iv Similarally (i) let $p(x) = 3x - 2$,

$$p(k) = 0 \implies 3k - 2 = 0 \implies 3k = 2 \implies k = \frac{2}{3}$$

So the zero of $p(x)$ is $\frac{2}{3}$.

v Similarally (i), let $p(x) = 3x$,

$$p(k) = 0 \implies 3k = 0 \implies k = 0$$

So 0 is the zero of $p(x)$.

2. POLYNOMIALS

vi Similarally (i), let $p(x) = ax, a \neq 0$,

$$p(k) = 0 \implies ak = 0 \implies k = 0$$

So 0 is zero of $p(x)$.

vii Similarally (i), let $p(x) = cx + d, c \neq 0, c, d$ are real numbers.

$$p(k) = 0 \implies ck + d = 0 \implies ck = -d \implies k = -\frac{d}{c}$$

So $-\frac{d}{c}$ is zero of $p(x)$.

2.7 Factorisation of Polynomials

In this section we discuss that when we have a polynomial $p(x)$ more than one degree then how can we write the polynomial $p(x)$ as product of two or more polynomials. This process is called factorisation of polynomials.

Theorem 2.7.1 Factor Theorem: If $p(x)$ is a polynamial of degree $n \geq 1$ and a is any real number, then (i) $x - a$ is a factor of $p(x)$, if $p(a) = 0$, and (ii) $p(a) = 0$, if $x - a$ is a factor of $p(x)$.

Proof. By remainder Theorem, for any polynomial $p(x) \exists q(x)$ such that $p(x) = (x - a)q(x) + p(a)$.

i If $p(a) = 0$, then $p(x) = (x - a)q(x)$, which show that $x - a$ is a factor of $p(x)$.

2. POLYNOMIALS

ii Since $x-a$ is a factor of $p(x)$, then $p(x) = (x-a)g(x)$ for some polynomial $g(x)$. So $p(a) = (a-a)g(a) = 0$.

□

Example 2.7.2. Examine whether $x+2$ is a factor of $x^3 + 3x^2 + 5x + 6$ and $2x + 4$.

Solution Now the zero of polynomial $x+2$ is -2, $p(x) = x+2 \implies -2+2 = 0$. We have polynomial $p(x) = x^3 + 3x^2 + 5x + 6$, then

$$p(-2) = (-2)^3 + 3(-2)^2 + 5(-2) + 6 = -8 + 12 - 10 + 6 = 18 - 18 = 0$$

So, by factor teheorem, $x+2$ is a factor of $x^3 + 3x^2 + 5x + 6$. Next consider $g(x) = 2x + 4$,

$$g(-2) = 2(-2) + 4 = -4 + 4 = 0$$

So, by factor teheorem, $x+2$ is a factor of $2x + 4$.

Example 2.7.3. Find the value of k, if $x-1$ is a factor of $4x^3 + 3x^2 - 4x + k$.

Solution If the monomial $x-1$ is a factor of $p(x) = 4x^3 + 3x^2 - 4x + k$, then by factor theorem $p(1) = 0$.

$$p(1) = 0 \implies 4(1)^3 + 3(1)^2 - 4(1) + k = 0 \implies 4 + 3 - 4 + k = 0 \implies k = -3$$

2. POLYNOMIALS

2.8 Solution of exercise 2.3

Problem 2.8.1. Find the remainder when $x^3 + 3x^2 + 3x + 1$ is divided by

i $x + 1$

ii $x - \frac{1}{2}$

iii x

iv $x + \pi$

v $5 + 2x$

Solution: Using Remainder Theorem

i We have $p(x) = x^3 + 3x^2 + 3x + 1$, putting $x = -1$. We get remainder

$$p(-1) = (-1)^3 + 3(-1)^2 + 3(-1) + 1 = -1 + 3 - 3 + 1 = 0$$

ii We have $p(x) = x^3 + 3x^2 + 3x + 1$, putting $x = \frac{1}{2}$. We get remainder

$$p\left(\frac{1}{2}\right) = \left(\frac{1}{2}\right)^3 + 3\left(\frac{1}{2}\right)^2 + 3\left(\frac{1}{2}\right) + 1 = \frac{1}{8} + 3\frac{1}{4} + 3\frac{1}{2} + 1 = \frac{63}{8}$$

iii We have $p(x) = x^3 + 3x^2 + 3x + 1$, putting $x = 0$. We get remainder

$$p(0) = (0)^3 + 3(0)^2 + 3(0) + 1 = 0 + 0 + 0 + 1 = 1$$

iv We have $p(x) = x^3 + 3x^2 + 3x + 1$, putting $x = -\pi$. We get remainder

$$p(-\pi) = (-1pi)^3 + 3(-\pi)^2 + 3(-\pi) + 1 = -\pi + 3\pi^2 - 3\pi + 1$$

2. POLYNOMIALS

v We have $p(x) = x^3 + 3x^2 + 3x + 1$, putting $x = -\frac{5}{2}$. We get remainder

$$p\left(-\frac{5}{2}\right) = \left(-\frac{5}{2}\right)^3 + 3\left(-\frac{5}{2}\right)^2 + 3\left(-\frac{5}{2}\right) + 1 = -\frac{125}{8} + 3\frac{25}{4} - 3\frac{5}{2} + 1 = -\frac{87}{8}$$

Problem 2.8.2. Find the remainder when $x^3 - ax^2 + 6x - a$ is divided by $x - a$.

Solution: We have polynomial $p(x) = x^3 - ax^2 + 6x - a$, when $p(x)$ is divided by $x - a$ then by Remainder Theorem, remainder get

$$p(a) = a^3 - aa^2 + 6a - a = a^3 - a^3 + 5a = 5a$$

Problem 2.8.3. Check whether $7 + 3x$ is a factor of $3x^3 + 7x$.

Solution: We have polynomial $p(x) = 3x^3 + 7x$, then by Remainder Theorem the remainder when $p(x)$ is divided by $7 + 3x$ is $p\left(-\frac{7}{3}\right)$.

$$p\left(-\frac{7}{3}\right) = 3\left(-\frac{7}{3}\right)^3 + 7\left(-\frac{7}{3}\right) = -\frac{490}{9}.$$

So by Factor Theorem the $7 + 3x$ is not factor of $p(x) = 3x^3 + 7x$ because remainder is not 0.

Factorisation of polynomial by splitting the middle term

Let quadratic is $ax^2 + bx + c$. To factorise the quadratic follow the following steps,

i Multiply the coefficien of x^2 and constant. Let $m = a \times c$.

2. POLYNOMIALS

ii Factorise the m as possible way. let possible factor of m are $a \times c, r \times s, p \times q$.

iii Consider that factor of m which sum is equal to the ceofficient of x, as $b = r + s$.

iv Now write the quadratic as $ax^2 + (r+s)x + c$.

After simplification we get factor of quadratic.

Example 2.8.4. Factorise $6x^2 + 17x + 5$ by splitting the middle term, and by using the Factor Theorem.

Solution 1: (By splitting method): The given quadratic is $6x^2 + 17x + 5$, now compairing to general form of quadratic polynomial we get $a = 6$, $b = 17$, $c = 5$. So $m = a \times c = 6 \times 5 = 30$. Now need find two numbers p, q such that $p + q = 17$.

$m = p \times q$	$p + q = b$	$p + q = 17$	choice
1×30	$1 + 30$	31	
2×15	$2 + 15$	17	✓
3×10	$3 + 10$	13	
5×6	$5 + 6$	11	

From the table,

$$6x^2 + 17x + 5 = 6x^2 + (2 + 15)x + 5$$

2. POLYNOMIALS

$$= 6x^2 + 2x + 15x + 5 = 2x(3x+1) + 5(3x+1)$$

$$= (3x+1)(2x+1)$$

Solution 2 (By using the Factor Theorem): We have $6x^2 + 17x + 5 = 6\left(x^2 + \frac{17}{6} + \frac{5}{6}\right) = 6p(x)$, say. If a and b are the zeroes of $p(x)$, then we can write $6x^2 + 17x + 5 = 6(x-a)(x-b) = 6[x^2 - (a+b)x + ab]$. So $ab = \frac{5}{6}$ and $a+b = -\frac{17}{6}$. Let us look at some posibilities for a and b.

a	b	$p(a)$	$p(b)$
$-\frac{1}{2}$	$\frac{1}{2}$	$p\left(-\frac{1}{2}\right) \neq 0$	$p\left(\frac{1}{2}\right) \neq 0$
$-\frac{1}{3}$	$\frac{1}{3}$	$p\left(-\frac{1}{3}\right) = 0, \checkmark$	$p\left(\frac{1}{3}\right) \neq 0$
$\frac{5}{3}$	$-\frac{5}{3}$	$p\left(\frac{5}{2}\right) \neq 0$	$p\left(-\frac{5}{2}\right) = 0, \checkmark$
$\frac{5}{2}$	$-\frac{5}{2}$	$p\left(\frac{5}{2}\right) \neq 0$	$p\left(-\frac{5}{2}\right) \neq 0$

Therefore $\left(x + \frac{1}{3}\right)$ and $\left(x + \frac{5}{2}\right)$ are factor of $p(x)$. So,

$$6x^2 + 17x + 5 = 6\left(x + \frac{1}{3}\right)\left(x + \frac{5}{2}\right) = 6\left(\frac{3x+1}{3}\right)\left(\frac{2x+5}{2}\right) = (3x+1)(2x+5).$$

Example 2.8.5. Factorise $y^2 - 5y + 6$ by using the Factor Theorem.

Solution: Let $p(y) = y^2 - 5y + 6$. Now, if $p(y) = (y-a)(y-b)$, you know that the constant term will be ab. So, $ab = 6$. Therefore, to determine of the factor $p(y)$, we determine the factor 6, because any one factor of 6 will be a and b. The factor of 6 are $1, 2$ and 3.

Now, $p(2) = 2^2 - 5 \times 2 + 6 = 4 - 10 + 6 = 0$.

So, $y - 2$ is a factor of $p(y)$.

2. POLYNOMIALS

Next $p(3) = 3^2 - 5 \times 3 + 6 = 9 - 15 + 6 = 0$.

So, $y - 3$ is also a factor of $p(y)$.

Therefore, $y^2 - 5y + 6 = (y-2)(y-3)$.

Example 2.8.6. Factorise $x^3 - 23x^2 + 142x - 120$.

Solution: Let $p(x) = x^3 - 23x^2 + 142x - 120$. If $p(x) = (x-a)(x-b)(x-c)$, then $abc = -120$, when coefficiend of x^3 is 1. Now we shall look all the factor of -120, whics are the $\pm 1, \pm 2, \pm 3, \pm 4, \pm 5, \pm 6, \pm 8, \pm 10, \pm 12, \pm 15, \pm 20, \pm 24, \pm 30, \pm 60$.

by trial metho, we find $p(1) = 1^3 - 23 \times 1^2 + 142 \times 1 - 120 = 1 - 23 + 142 - 120 = 0$.

So, $x - 1$ is factor of $p(x)$.

Similarally $p(10) = 10^3 - 23 \times 10^2 + 142 \times 10 - 120 = 1000 - 2300 + 1420 - 120 = 0$.

So, $x - 10$ will be the factor of $p(x)$.

Similarally $p(12) = 12^2 - 23 \times 12^2 + 142 \times 12 - 120 = 1728 - 3312 + 1704 - 120 = 0$.

So, $x - 12$ will be the factor of $p(x)$. Hence $p(x) = (x-1)(x-10)(x-12)$.

2.9 Solution of Exercise 2.4

Problem 2.9.1. Determine which of the following polynomial has $(x+1)$ factor:

i $x^3 + x^2 + x + 1$

ii $x^4 + x^3 + x^2 + x + 1$

2. POLYNOMIALS

iii $x^4 + 3x^3 + 3x^2 + x + 1$

iv $x^3 - x^2 - (2 + \sqrt{2})x + \sqrt{2}$

Solution: If $x + 1$ is the factor of a polynomial $p(x)$, then $p(-1) = 0$.

i Let $p(x) = x^3 + x^2 + x + 1$ Now,

$$p(-1) = (-1)^3 + (-1)^2 + (-1) + 1 = -1 + 1 - 1 + 1 = 0.$$

So, $x + 1$ is the factor of $p(x)$.

ii Let $p(x) = x^4 + x^3 + x^2 + x + 1.$

Then,

$$p(-1) = (-1)^4 + (-1)^3 + (-1)^2 + (-1) + 1 = 1 - 1 + 1 - 1 + 1 = 1.$$

So, $x + 1$ is not the factor of $x^3 + x^2 + x + 1$.

iii Let $p(x) = x^4 + 3x^3 + 3x^2 + x + 1.$

Then,

$$p(-1) = (-1)^4 + 3(-1)^3 + 3(-1)^2 + (-1) + 1 = 1 - 3 + 3 - 1 + 1 = 1.$$

So, $x + 1$ is not the factor of $p(x) = x^4 + 3x^3 + 3x^2 + x + 1$.

iv Let $p(x) = x^3 - x^2 - (2 + \sqrt{2})x + \sqrt{2}.$

Then,

$$p(-1) = (-1)^3 - (-1)^2 - \left(2 + \sqrt{2}\right)(-1) + \sqrt{2} = -1 - 1 + \left(2 + \sqrt{2}\right) + \sqrt{2}$$

2. POLYNOMIALS

$$= 2\sqrt{2}.$$

So, $x + 1$ is not factor of the $p(x) = x^3 - x^2 - (2 + \sqrt{2})x + \sqrt{2}$.

Problem 2.9.2. Use Factor Theorem to determine whether $g(x)$ is a factor of $p(x)$ in each of the following cases:

i $p(x) = 2x^3 + x^2 - 2x - 1$, $g(x) = x + 1$

ii $p(x) = x^3 + 3x^2 + 3x + 1$, $g(x) = x + 2$

iii $p(x) = x^3 - 4x^2 + x + 6$, $g(x) = x - 3$

Solution:

i Let $p(x) = 2x^3 + x^2 - 2x - 1$, and $g(x) = x + 1$, then

$$p(-1) = 2(-)^3 + (-1)^2 - 2(-1) - 1 = -2 + 1 + 2 - 1 = 0.$$

So, by Factor Theorem $g(x) = x + 1$ is a factor of $p(x)$.

ii Let $p(x) = x^3 + 3x^2 + 3x + 1$ and $g(x) = x + 2$ then

$$p(-2) = (-2)^3 + 3(-2)^2 + 3(-2) + 1 = -8 + 3 \times 4 - 6 + 1$$

$$= -8 + 12 - 6 + 1 = -1.$$

So, by Factor Theorem $g(x)$ is not factor of $p(x)$.

iii $p(x) = x^3 - 4x^2 + x + 6$ and $g(x) = x - 3$, then

$$p(3) = 3^3 - 4 \times 3^2 + 3 + 6 = 27 - 36 + 3 + 6 = 0.$$

So, by Factor Theorem $g(x)$ is a factor of $p(x)$.

Problem 2.9.3. Find the value of k, if $x - 1$ is a factor of $p(x)$ in each of the following cases:

i $p(x) = x^2 + x + k$

ii $p(x) = 2x^2 + kx + \sqrt{2}$

iii $p(x) = kx^2 - \sqrt{2}x + 1$

iv $p(x) = kx^2 - 3x + k$

Solution:

i Let $p(x) = x^2 + x + k$, if $x - 1$ is a factor of $p(x)$. Then,

$$p(1) = 0 \implies 1^2 + 1 + k = 0 \implies 2 + k = 0 \implies k = -2.$$

ii Let $p(x) = 2x^2 + kx + \sqrt{2}$, if $x - 1$ is a factor of $p(x)$. Then,

$$p(1) = 0 \implies 2.1^2 + k1 + \sqrt{2} = 0 \implies 2 + k + \sqrt{2} = 0 \implies k = -2 - \sqrt{2}.$$

iii Let $p(x) = kx^2 - \sqrt{2}x + 1$, if $x - 1$ is a factor of $p(x)$. Then,

$$p(1) = 0 \implies k.1 - \sqrt{2}.1 + 1 = k - \sqrt{2} + 1 = 0 \implies k = -1 + \sqrt{2}.$$

2. POLYNOMIALS

iv $p(x) = kx^2 - 3x + k$, if $x - 1$ is a factor of $p(x)$. Then,

$$p(1) = 0 \implies k.1^2 - 3.1 + k = 0 \implies k - 3 + 1 = 0 \implies 2k = 3 \implies k = \frac{3}{2}.$$

Problem 2.9.4. Factorise:

i $12x^2 - 7x + 1$

ii $2x^2 + 7x + 3$

iii $6x^2 + 5x - 6$

iv $3x^2 - x - 4$

Solution:

i Let $12x^2 - 7x + 1$ is given polynomial, then

$$12x^2 - 7x + 1 = 12x^2 - (4+3)x + 1$$

$$= 12x^2 - 4x - 3x + 1 = 4x(3x - 1) - 1(3x - 1)$$

$$= (3x - 1)(4x - 1).$$

ii Let $2x^2 + 7x + 3$ is a given polynomial, then

$$2x^2 + 7x + 3 = 2x^2 + (6+1)x + 3$$

$$= 2x^2 + 6x + x + 3 = 2x(x + 3) + 1(x + 3) = (x + 3)(2x + 1).$$

iii Let $6x^2 + 5x - 6$ is a given polynomial, then

$$6x^2 + 5x - 6 = 6x^2 + (9-4)x - 6$$

$$= 6x^2 + 9x - 4x - 6 = 3x(2x+3) - 2(3x+2)$$

$$= (2x+3)(3x-2).$$

iv Let $3x^2 - x - 4$ is a polynomial, then

$$3x^2 - x - 4 = 3x^2 + (3-4)x - 4$$

$$= 3x^2 + 3x - 4x - 4 = 3x(x+1) - 4(x+1)$$

$$= (x+1)(3x-4).$$

Problem 2.9.5. Factorise :

i $x^3 - 2x^2 - x + 2$

ii $x^3 - 3x^2 - 9x - 5$

iii $x^3 + 13x^2 + 32x + 20$

iv $2y^3 + y^2 - 2y - 1$

Solution:

2. POLYNOMIALS

i Let $x^3 - 2x^2 - x + 2$ is a polynomial. The constant term of polynamial is 2, now the factor of constant term are $\pm 1, \pm 2$. The value of polynomial at factor of constant term,

$$p(1) = 1^3 - 2 \times 1^2 - 1 + 2 = 1 - 2 - 1 + 2 = 0.$$

So, by Factor Theorem $x - 1$ is factor of $x^3 - 2x^2 - x + 2$. Hence

$$x^3 - 2x^2 - x + 2 = x^2(x-1) - x(x-1) - 2(x-1) = (x-1)(x^2 - x - 2)$$

$$= (x-1)\{x^2 - 2x + x - 2\} = (x-1)\{x(x-2) + (x-2)\}$$

$$= (x-1)(x-2)(x+1).$$

ii Let $p(x) = x^3 - 3x^2 - 9x - 5$. Now the factors of constant term of $p(x)$ are $\pm 1, \pm 5$. Next

$$p(5) = 5^3 - 3 \times 5^2 - 9 \times 5 - 5 = 125 - 75 - 45 - 5 = 0.$$

So, by Factor Theorem $x - 5$ is the factor of $p(x)$. Hence

$$x^3 - 3x^2 - 9x - 5 = x^2(x-5) + 2x^2 - 9x - 5 = x^2(x-5) + 2x(x-5) + x - 5$$

$$= (x-5)(x^2 + 2x + 1) = (x-5)(x+1)(x+1).$$

iii Let $p(x) = x^3 + 13x^2 + 32x + 20$. The factors of constant term of $p(x)$ are $\pm 2, \pm 5$. Now

$$p(-2) = (-2)^3 + 13 \times (-2)^2 + 32(-2) + 20 = -8 + 52 - 64 + 20 = 0.$$

2. POLYNOMIALS

So, by Factor Theorem $x + 2$ is the factor of $p(x)$. Hence

$$p(x) = x^3 + 13x^2 + 32x + 20 = x^2(x+2) + 11x^2(x+2) + 10(x+2)$$

$$= (x+2)(x^2+11x+10) = (x+2)(x^2+x+10x+10) = (x+2)(x+1)(x+10).$$

iv Let $p(y) = 2y^3 + y^2 - 2y - 1$. The factor of constant term are ± 1. Then

$$p(1) = 2.1^3 + 1^2 - 2 \times 1 - 1 = 2 + 1 - 2 - 1 = 0.$$

So, by Factor Theorem $y - 1$ is the factor of $p(y)$. Hence

$$2y^3 + y^2 - 2y - 1 = 2y^2(y-1) + 3y(y-1) + (y-1) = (y-1)(2y^2 + 3y + 1)$$

$$= (y-1)(2y^2 + 2y + y + 1) = (y-1)(y+1)(2y+1).$$

2.10 Algebraic Identities

Algebraic Identity : An algebraic identity is an algebraic equation that is true for all values of the variables occurring in it.

There are some algebraic identities,

Identity I : $(x+y)^2 = x^2 + 2xy + y^2$

Identity II : $(x-y)^2 = x^2 - 2xy + y^2$

Identity III : $x^2 - y^2 = (x+y)(x-y)$

Identity IV : $(x+a)(x+b) = x^2 + (a+b)x + ab$

2. POLYNOMIALS

Example 2.10.1. Find the following product using appropriate identities:

i $(x+3)(x+3)$

ii $(x-3)(x+5)$

Solution :

i We have identity I : $(x+y)^2 = x^2 + 2xy + y^2$. Putting $y = 3$ in identity.

We get

$$(x+3)(x+3) = (x+3)^2 = x^2 + 2(x)(3) + 3^2 = x^2 + 3x + 9.$$

ii Using identity IV : $(x+a)(x+b) = x^2 + (a+b)x + ab$, we get

$$(x-3)(x+5) = x^2 + (-3+5)x + (-3)(5) = x^2 + 2x - 15.$$

Example 2.10.2. Evaluate 105×106 without multiplying directly.

Solution :

$$105 \times 106 = (100 = 5)(100 + 6)$$

$$= (100)^2 + (5+6)100 + 5 \times 6 = 10000 + 1100 + 30$$

$$= 11130.$$

Example 2.10.3. Factorise :

i $49a^2 + 70ab + 25b^2$

ii $\frac{25}{4}x^2 - \frac{y^2}{9}$

Solution : Now we can write the

i $49a^2 + 70ab + 25b^2$, as

$$49a^2 + 70ab + 25b^2 = (7a)^2 + 2(7a)(5b) + (5b)^2$$

$$= (7a + 5b)^2 = (7a + 5b)(7a + 5b)$$

ii We have $\frac{25}{4}x^2 - \frac{y^2}{9}$. It can be write

$$\frac{25}{4}x^2 - \frac{y^2}{9} = \left(\frac{5}{2}x\right)^2 - \left(\frac{y}{3}\right)^2$$

Using the identity III, we get

$$\left(\frac{5}{2}x\right)^2 - \left(\frac{y}{3}\right)^2 = \left(\frac{5}{2}x - \frac{y}{3}\right)\left(\frac{5}{2}x - \frac{y}{3}\right)$$

Identity V : $(x + y + z)^2 = x^2 + y^2 + z^2 + 2xy + 2yz + 2zx$

Identity VI : $(x + y)^3 = x^3 + y^3 + 3xy(x + y) = x^3 + y^3 + 3x^2y + 3xy^2$

Identity VII : $(x - y)^3 = x^3 - y^3 - 3xy(x - y) = x^3 - y^3 = 3x^2y + 3xy^2$

Example 2.10.4. Write $(3a + 4b + 5c)^2$ in expanded form.

Solution : Comparing the given expression $(3a+4b+5c)^2$ with $(x+y+z)^2$, we find that

$$x = 3a, y = 4b, z = 5c.$$

2. POLYNOMIALS

So, using Identiti V, we get

$$(3a + 4b + 5c)^2 = (3a)^2 + (4b)^2 + (5c)^2 + 2(3a)(4b) + 2(4b)(5c) + 2(5c)(3a)$$

$$= 9a^2 + 16b^2 + 25c^2 + 24ab + 40bc + 30ca$$

Example 2.10.5. Expand $(4a - 2b - 3c)^2$.

   Solution : We have $(4a - 2b - 3c)^2$, now using Identity V,

$$(4a - 2b - 3c)^2 = [4a + (-2b) + (-3c)]^2$$

$$= (4a)^2 + (-2b)^2 + (-3c)^2 + 2(4a)(-2b) + 2(-2b)(-3c) + 2(-3c)(4a)$$

$$= 16a^2 + 4b^2 + 9c^2 - 16ab + 12bc - 24ac$$

Example 2.10.6. Factorise $4x^2 + y^2 + z^2 - 4xy - 2yz + 4xz$.

   Solution : We have expression $4x^2 + y^2 + z^2 - 4xy - 2yz + 4xz$. Now it can be express as

$$4x^2 + y^2 + z^2 - 4xy - 2yz + 4xz = (2x)^2 + (-y)^2 + (z)^2 + 2(2x)(-y) + 2(-y)(z) + 2(2x)(z)$$

using identity V

$$= [2x - y + z]^2 = (2x - y + z)(2x - y + z)$$

Example 2.10.7. Wirte the following cubes in the expanded form:

i $(3a + 4b)^3$

2. POLYNOMIALS

ii $(5a - 3b)^3$

Solution :

i Using identity VI

$$(3a + 4b)^3 = (3a)^3 + (4b)^3 + 3(3a)^2(4b) + 3(3a)(4b)^2 = 27a^3 + 64b^3 + 108a^2b + 144ab^2.$$

ii Using identity VI

$$(5a - 3b)^3 = (5a^3) + (-3b)^3 + 3(5a)^2(-3b) + 3(5a)(-3b)^2 = 125a^3 - 27b^3 - 225a^2b + 135ab^2.$$

Use of Identity for numerical calculation

Example 2.10.8. Evaluate each of the following using sutable identies:

i $(104)^3$

ii $(999)^3$

Solution:

i We have

$$(104)^3 = (100 + 4)^3$$

$$= 100^3 + 4^3 + 3 \times 100^2 \times 4 + 3 \times 100 \times 4^2$$

$$= 1000000 + 16 + 12 \times 10000 + 48 \times 100$$

2. POLYNOMIALS

$$= 1000000 + 16 + 120000 + 4800 = 101204816.$$

ii We have

$$(999)^3 = (1000-1)^3$$

$$= 1000^3 - 1^3 + 3 \times 1000^2 \times (-1) + 3 \times 1000 \times (-1)^2$$

$$= 1000000000 - 1 - 3 \times 1000000 + 3 \times 1000$$

$$= 1000000000 - 1 - 3000000 + 3000 = 1000003000 - 3000001$$

$$= 99702999.$$

Example 2.10.9. Factorise $8x^3 + 27y^3 + 36x^2 + 54xy^2$

Solution: The given expression can be written as,

$$8x^3 + 27y^3 + 36x^2 + 54xy^2 = (2x)^3 + (3y)^3 + 3(2x)^2(3y) + 3(2x)(3y^2)$$

$$= (2x + 3y)^3 = (2x + 3y)(2x + 3y)(2x + 3y)$$

Identity VIII:

$$x^3 + y^3 + z^3 - 3xyz = (x + y + z)(x^2 + y^2 + x^2 - xy - yz - zx)$$

Example 2.10.10. Factorise : $8x^3 + y^3 + 27z^3 - 18xyz$

Solution : We have

$$8x^3 + y^3 + 27z^3 - 18xyz = (2x)^3 + (y^3) + (3z)^3 - 3(2x)(y)(3z)$$

2. POLYNOMIALS

$$= (2x + y + 3z)\left[(2x)^2 + y^2 + (3z)^2 - (2x)y + y(3z) - (3z)(2x)\right]$$

$$= (2x + y + 3z)(4x^2 + y^2 + 9z^2 - 2xy - 3yz - 6zx)$$

using identity VIII

2.11 Solution of Exercise 2.5

Problem 2.11.1. Use sutaible identities to find the following products:

i $(x+4)(x+10)$

ii $(x+8)(x-10)$

iii $(3x+4)(3x-5)$

iv $\left(y^2 + \frac{3}{2}\right)\left(y^2 - \frac{3}{2}\right)$

v $(3-2x)(3+2x)$

Solution:

i Using identities IV, we get

$$(x+4)(x+10) = x^2 + (4+10)x + 4 \times 10 = x^2 + 14x + 40$$

ii Using identities IV, we get

$$(x+8)(x-10) = x^2 + (8-10)x + 8 \times (-10) = x^2 - 2x - 80$$

2. POLYNOMIALS

iii Using identities IV, we get

$$(3x+4)(3x-5) = (3x)^2 + (4-5)(3x) + 4(-5) = 9x^2 - 3x - 20$$

iv Using identities III, we get

$$\left(y^2+\frac{3}{2}\right)\left(y^2-\frac{3}{2}\right) = (y^2)^2 - \left(\frac{3}{2}\right)^2 = y^4 - \frac{9}{4}$$

v Using identities III, we get

$$(3-2x)(3+2x) = 3^2 - (2x)^2 = 9 - 4x^2$$

Problem 2.11.2. Evaluate the following products without multiplying directly:

i 103×107

ii 95×96

iii 104×96

Solution: Using Identity IV

i It cab be written as $103 \times 107 = (100+3)(100+7)$

$$(100+3)(100+7) = 100^2 + (3+7)100 + 3 \times 7 = 10000 + 10 \times 100 + 21$$

$$= 10000 + 1000 + 21 = 11021$$

2. POLYNOMIALS

ii It cab be written as $95 \times 96 = (100-5)(100-4)$

$$(100-5)(100-4) = 100^2 + (-5-4)100 + (-5)(-4) = 10000 - 9 \times 100 + 20$$

$$= 10000 - 900 + 20 = 9120$$

iii It cab be written as $104 \times 96 = (100+4)(100-4)$ Using Identitiy III

$$(100+4)(100-4) = 100^2 - 4^2 = 10000 - 16 = 9984$$

Problem 2.11.3. Factorise the following using appropriate identities:

i $9x^2 + 6xy + y^2$

ii $4y^2 - 4y + 1$

iii $x^2 - \frac{y^2}{100}$

Solution:

i It can be written as

$$9x^2 + 6xy + y^2 = (3x)^2 + 2 \times 3x \times y + y^2$$

Using Identity I

$$(3x)^2 + 2 \times 3x \times y + y^2 = (3x+y)^2 = (3x+y)(3x+y)$$

2. POLYNOMIALS

ii It can be written as

$$4y^2 - 4y + 1 = (2y)^2 - 2 \times 2y \times 1 + 1^2$$

Using Identity II

$$(2x)^2 - 2 \times 2x \times 1 + 1^2 = (2y-1)^2 = (2y-1)(2y-1).$$

iii It can be written as

$$x^2 - \frac{y^2}{100} = (x)^2 - \left(\frac{y}{10}\right)^2$$

Using Identity III

$$(x)^2 - \left(\frac{y}{10}\right)^2 = \left(x + \frac{y}{10}\right)\left(x - \frac{y}{10}\right)$$

Problem 2.11.4. Expand each of the following, using suitable identities:

i $(x + 2y + 4z)^2$

ii $(2x - y + z)^2$

iii $(-2x + 3y + 2z)^2$

iv $(3a - 7b - c)^2$

v $(-2x + 5y - 3z)^2$

vi $\left[\frac{1}{4}a - \frac{1}{2}b + 1\right]^2$

Solution: Using the Identity V

i The given expression $(x + 2y + 4z)^2$, so

$$(x + 2y + 4z)^2 = x^2 + (2y)^2 + (4z)^2 + 2x(2y) + 2(2y)(4z) + 2(4z)x$$

$$= x^2 + 4y^2 + 16z^2 + 4xy + 16yz + 4zx$$

ii The given expression $(2x - y + z)^2$, so

$$(2x - y + z)^2 = (2x)^2 + (-y)^2 + z^2 + 2(2x)(-y) + 2(-y)z + 2z(2x)$$

$$= 4x^2 + y^2 + z^2 - 4xy - 2yz + 2zx$$

iii The given expression $(-2x + 3y + 2z)^2$, so

$$(-2x+3y+2z)^2 = (-2x)^2 + (3y)^2 + (2z)^2 + 2(-2x)(3y) + 2(3y)(2z) + 2(2z)(-2x)$$

$$= 4x^2 + 9y^2 + 4z^2 - 12xy + 12yz - 8zx$$

iv The given expression $(3a - 7b - c)^2$, so

$$(3a-7b-c)^2 = (3a)^2 + (-7b)^2 + (-c)^2 + 2(3a)(-7b) + 2(-7b)(-c) + 2(-c)(3a)$$

$$= 9a^2 + 49b^2 + c^2 - 42ab + 14bc - 6ca$$

v The given expression $(-2x + 5y - 3z)^2$, so

$$(-2x+5y-3z)^2 = (-2x)^2 + (5y)^2 + (-3z)^2 + 2(-2x)(5y) + 2(5y)(-3z) + 2(-3z)(-2x)$$

$$= 4x^2 + 25y^2 + 9z^2 - 20xy - 30yz + 12zx$$

2. POLYNOMIALS

vi The given expression $\left[\frac{1}{4}a - \frac{1}{2}b + 1\right]^2$, so

$$\left[\frac{1}{4}a - \frac{1}{2}b + 1\right]^2 = \left(\frac{1}{4}a\right)^2 + \left(-\frac{1}{2}b\right)^2 + 1^2 + 2\left(\frac{1}{4}a\right)\left(-\frac{1}{2}b\right) + 2\left(-\frac{1}{2}b\right) \times 1 + 2 \times 1\left(\frac{1}{4}a\right)$$

$$= \frac{1}{16}a^2 + \frac{1}{4}b^2 + 1 - \frac{1}{4}ab - b + \frac{1}{2}a$$

Problem 2.11.5. Factorise:

i $4x^2 + 9y^2 + 16z^2 + 12xy - 24yz - 16zx$

ii $2x^2 + y^2 + 8z^2 - 2\sqrt{2}xy + 4\sqrt{2}yz - 8zx$

Solution: Using identity V

i We have expression $4x^2 + 9y^2 + 16z^2 + 12xy - 24yz - 16zx$. It can be written as

$$4x^2 + 9y^2 + 16z^2 + 12xy - 24yz - 16zx$$

$$= (2x)^2 + (3y)^2 + (-4z)^2 + 2(2x)(3y) + 2(3y)(-4z) + 2(-4z)(2x)$$

$$= (2x + 3y - 4z)^2 = (2x + 3y - 4z)(2x + 3y - 4z)$$

ii We have expression $2x^2 + y^2 + 8z^2 - 2\sqrt{2}xy + 4\sqrt{2}yz - 8zx$. So it can be written as

$$2x^2 + y^2 + 8z^2 - 2\sqrt{2}xy + 4\sqrt{2}yz - 8zx$$

$$= \left(-\sqrt{2}x\right)^2 + y^2 + \left(2\sqrt{2}z\right)^2 + 2\left(-\sqrt{2}x\right)y + 2y\left(2\sqrt{2}z\right) + 2\left(2\sqrt{2}z\right)\left(-\sqrt{2}x\right)$$

$$= \left(-\sqrt{2}x + y + 2\sqrt{2}z\right)^2 = \left(-\sqrt{2}x + y + 2\sqrt{2}z\right)\left(-\sqrt{2}x + y + 2\sqrt{2}z\right)$$

2. POLYNOMIALS

Problem 2.11.6. Write the following cubes in expanded form:

i $(2x+1)^3$

ii $(2a-3b)^3$

iii $\left[\frac{3}{2}x+1\right]^3$

iv $\left[x-\frac{2}{3}y\right]^3$

Solution: Using the Idenditi VI for i and iii, and identity VII for ii and iv,

i We have $(2x+1)^3$, so

$$(2x+1)^3 = (2x)^3 + 1^3 + 3(2x)^2 \times 1 + 3(2x)1^2 = 8x^3 + 1 + 12x^2 + 6x$$

ii We have $(2a-3b)^3$, so

$$(2a-3b)^3 = (2a)^3 - (3b)^3 - 3(2a)^2(3b) + 3(2a)(3b)^2$$

$$= 8a^3 - 27b^3 - 36a^2b + 54ab^2$$

iii We have $\left[\frac{3}{2}x+1\right]^3$, so

$$\left[\frac{3}{2}x+1\right]^3 = \left(\frac{3}{2}x\right)^3 + 1^3 + 3\left(\frac{3}{2}x\right)^2 \times 1 + 3\left(\frac{3}{2}x\right)1^2$$

$$= \frac{27}{8}x^3 + 1 + \frac{81}{4}x^2 + \frac{9}{2}x$$

2. POLYNOMIALS

iv We have $\left[x - \frac{2}{3}y\right]^3$, so

$$\left[x - \frac{2}{3}y\right]^3 = x^3 - \left(\frac{2}{3}y\right)^3 - 3x^2\left(\frac{2}{3}y\right) + 3x\left(\frac{2}{3}y\right)^2$$

$$= x^3 - \frac{8}{27}y^3 - 2x^2y + \frac{4}{3}y^2x$$

Problem 2.11.7. Evaluate the following using sutable identites:

i $(99)^3$

ii $(102)^3$

iii $(998)^3$

Solution: To evaluate the problem i and iii apply the identities VII and for ii apply identities VI.

i It can be express as $(99)^3 = (100 - 1)^3$. So,

$$(100 - 1)^3 = 100^3 - 1^3 - 3 \times 100^2 \times 1 + 3 \times 100 \times 1$$

$$1000000 - 1 - 30000 + 300 = 970299.$$

ii It can be express as $(102)^3 = (100 + 2)^3$. So,

$$(100 + 2)^3 - 100^3 + 2^3 + 3 \times 100^2 \times 2 + 3 \times 100 \times 2^2$$

$$= 1000000 + 8 + 3 \times 10000 \times 2 + 3 \times 100 \times 4 = 1000000 + 8 + 120000 + 1200$$

$$= 1121208$$

2. POLYNOMIALS

iii It can be express as $(998)^3 = (1000-2)^3$. So,

$$(1000-2)^3 = 1000^3 - 2^3 - 3 \times 1000^2 \times 2 + 3 \times 1000 \times 2^2$$

$$= 100000000 - 8 - 6 \times 1000000 + 12000 = 994011992$$

Problem 2.11.8. Factorise each of the following:

i $8a^3 + b^3 + 12a^2b + 6ab^2$

ii $8a^3 - b^3 - 12a^2b + 6ab^2$

iii $27 - 125a^3 - 135a + 225a^2$

iv $64a^3 - 27b^3 - 144a^2b + 108ab^2$

v $27p^3 - \frac{1}{216} - \frac{9}{2}p^2 + \frac{1}{4}p$

Solution: Apply the identity VI for i and VII for (ii, iii, iv,v)

i We have expression $8a^3 + b^3 + 12a^2b + 6ab^2$.

$$8a^3 + b^3 + 12a^2b + 6ab^2 = (2a)^3 + b^3 + 3(2a)^2(b) + 3(2a)(b^2)$$

$$= (2a+b)^3 = (2a+b)(2a+b)(2a+b)$$

ii We have expression $8a^3 - b^3 - 12a^2b + 6ab^2$. So,

$$8a^3 - b^3 - 12a^2b + 6ab^2 = (2a)^3 - b^3 - 3(2a)^2(b) + 3(2a)(b^2)$$

$$= (2a-b)^3 = (2a-b)(2a-b)(2a-b)$$

2. POLYNOMIALS

iii We have expression $27 - 125a^3 - 135a + 225a^2$. So,

$$27 - 125a^3 - 135a + 225a^2 = 3^3 - (5a)^3 + 3(5a)^2 3 - 3(5a)3^2$$

$$= (3 - 5a)^3 = (3 - 5a)(3 - 5a)(3 - 5a)$$

iv We have expression $64a^3 - 27b^3 - 144a^2b + 108ab^2$. So,

$$64a^3 - 27b^3 - 144a^2b + 108ab^2 = (4a)^3 - (3b)^3 - 3(4a)^2(3b) + 3(4a)(3b)^2$$

$$= (4a - 3b)^3 = (4a - 3b)(4a - 3b)(4a - 3b).$$

v We have expression $27p^3 - \frac{1}{216} - \frac{9}{2}p^2 + \frac{1}{4}p$. So,

$$27p^3 - \frac{1}{216} - \frac{9}{2}p^2 + \frac{1}{4}p = (3p)^3 - \left(\frac{1}{6}\right)^3 - 3(3p)^2 \frac{1}{6} + 3\left(\frac{1}{6}\right)^2 (3p)$$

$$= \left(3p - \frac{1}{6}\right)^3 = \left(3p - \frac{1}{6}\right)\left(3p - \frac{1}{6}\right)\left(3p - \frac{1}{6}\right).$$

Problem 2.11.9. Verify:

i $x^3 + y^3 = (x + y)(x^2 - xy + y^2)$

ii $x^3 - y^3 = (x - y)(x^2 + xy + y^2)$

Solution:

i From equation i

$$RHS = (x + y)(x^2 - xy + y^2) = x^3 - x^2y + xy^2 + yx^2 - xy^2 + y^3$$

$$= x^3 + y^3 = LHS$$

ii From equation ii

$$RHS = (x-y)(x^2 + xy + y^2) = x^3 + x^2y + xy^2 - yx^2 - xy^2 - y^3$$

$$= x^3 - y^3 = LHS$$

Problem 2.11.10. Factorise each of the following:

i $27y^3 + 125z^3$

ii $64m^3 - 343n^3$

Solution:

i we have $27y^3 + 125z^3$. Using identity VI

$$27y^3 + 125z^3 = (3y + 5z)\left[(3y)^2 - (3y)(5z) + (5z)^2\right]$$

$$= (3y + 5z)(9y^2 - 15yz + 25z^2)$$

ii We have $64m^3 - 343n^3$. Using identity VII

$$64m^3 - 343n^3 = (4m - 7n)\left[(4m)^2 + (4m)(7n) + (7n)^2\right]$$

$$= (4m - 7n)(16m^2 + 28mn + 49n^2).$$

2. POLYNOMIALS

Problem 2.11.11. Factorise $27x^3 + y^3 + z^3 - 9xyz$.

Solution: Apply identity VIII, it can be express as

$$27x^3 + y^3 + z^3 - 9xyz = (3x + y + z)\left((3x)^2 + y^2 + z^2 - (3x)y - yz - z(3x)\right)$$

$$= (3x + y + z)(9x^2 + y^2 + z^2 - 3xy - yz - 3zx)$$

Problem 2.11.12. Verify

$$x^3 + y^3 + z^3 - 3xyz = \frac{1}{2}(x + y + z)\left[(x - y)^2 + (y - z)^2 + (z - x)^2\right]$$

Solution:

$$RHS = \frac{1}{2}(x + y + z)\left[(x - y)^2 + (y - z)^2 + (z - x)^2\right]$$

$$= \frac{1}{2}(x + y + z)\left[x^2 - 2xy + y^2 + y^2 - 2yz + z^2 + z^2 - 2zx + x^2\right]$$

$$= \frac{1}{2}(x + y + z)(2x^2 + 2y^2 + 2z^2 - 2xy - 2yz - 2zx)$$

$$= \frac{1}{2} \times 2(x + y + z)(x^2 + y^2 + z^2 - xy - yz - zx)$$

$$= (x + y + z)(x^2 + y^2 + z^2 - xy - yz - zx)$$

$$= x^3 + y^3 + z^3 - 3xyz = LHS$$

Problem 2.11.13. If $x + y + z = 0$, show that $x^3 + y^3 + z^3 - 3xyz$.

Solution: From identity VIII,

$$x^3 + y^3 + z^3 - 3xyz = (x + y + z)(x^2 + y^2 + z^2 - xy - yz - zx)$$

, putting $x + y + z = 0$, We get

$$x^3 + y^3 + z^3 - 3xyz = 0 \times (x^2 + y^2 + z^2 - xy - yz - zx)$$

$$x^3 + y^3 + z^3 - 3xyz = 0$$

$$x^3 + y^3 + z^3 = 3xyz.$$

Problem 2.11.14. Without actually calculating the cubes, find the value of each of the following:

i $(-12)^3 + (7)^3 + (5)^3$

ii $(28)^3 + (-15)^3 + (-13)^3$

Solution:

i Here given expression $(-12)^3 + (7)^3 + (5)^3$ and $-12 + 5 + 7 = 0$, So use the result of problem (2.11.14), we get

$$(-12)^3 + (7)^3 + (5)^3 = 3 \times (-12) \times 5 \times 7 = -1260$$

ii Here given expression $(28)^3 + (-15)^3 + (-13)^3$ and $28 - 15 - 13 = 0$, So use the result of problem 2.11.14, we get

$$(28)^3 + (-15)^3 + (-13)^3 = 3 \times 28(-15)(-13) = 16380.$$

Problem 2.11.15. Give possible expressions for the length and breadth of each of the following rectangles, in which their area are given:

2. POLYNOMIALS

i Area: $25a^2 - 35a + 12$

ii Area: $35y^2 + 13y - 12$

Solution: The area of recatngle is the product of length and breadth, then two possible factor of area of rectangle will be length and breadth of given rectangle,

i Area of rectangle is $25a^2 - 36a + 12$. So,

$$25a^2 - 35a + 12 = 25a^2 - 20a - 15a + 12 = 5a(5a - 4) - 3(5a - 4)$$

$$= (5a - 4)(5a - 3)$$

Hence, Lenth=$(5a - 4)$ and Breadth=$(5a - 3)$

ii Area of rectangle is $35y^2 + 13y - 12$. So,

$$35y^2 + 13y - 12 = 35y^2 + 28y - 15y - 12 = 7y(5y + 4) - 3(5y + 4)$$

$$= (5y + 4)(7y - 3)$$

Hence, Length=$(5y + 4)$ and Breadth=$(7y - 3)$

Problem 2.11.16. What is the possible expressions for the dimensions of the cuboids whose volumes are given below ?

i Volume: $3x^2 - 12x$

2. POLYNOMIALS

ii Volume: $12ky^2 + 8ky - 20k$

Solution: The volume of cuboid $V = lbh$, So possible three factor of volume expression will be lenth, breadth and height of cuboid.

i Volume is $3x^2 - 12x$. So,

$$3x^2 - 12x = 3x(x - 4)$$

. Hence Length=3, Breadth=x and Height=$(x - 4)$.

ii Volume is $12ky^2 + 8ky - 20k$. So,

$$12ky^2 + 8ky - 20k = 4k(3y^3 + 2y - 5) = 4k(3y^2 + 5y - 3y - 5)$$

$$= 4k\,[y(3y + 5) - (3y + 5)] = 4k(3y + 5)(y - 1)$$

Hence Length=$4k$, Breadth=$(3y + 5)$ and Height=$(y - 1)$.

3 COORDINATE GEOMETRY

3.1 Introduction

In number system you have seen that each real number can be placed a unique place on real line.

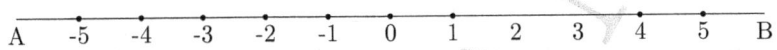

Figure 3.1.1: Positioning of point

But when we talk about position of a point in a plane then you can not decide the position of point by only one number line.

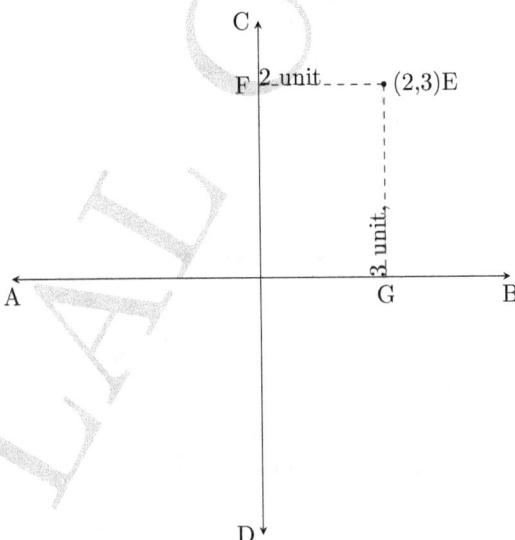

Figure 3.1.2: Positioning of point

3. COORDINATE GEOMETRY   90

Consider a point $E(2,3)$, how to find the correct position of 'E' ? We need two perpendicular lines AB and CD. The position of point E is 2 unit fram line CD and 3 unit from AB, that is exact position of E.

3.2 Solution of Exercise 3.1

Problem 3.2.1. How you discribe the position of lamp on your study table to another person?

Solution: Frist we name the side of table, then help of any two perpendicular and adjacent side of table.

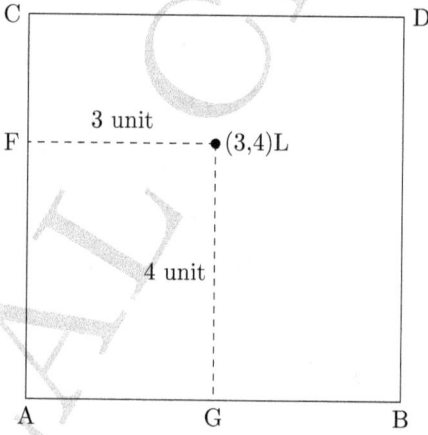

Figure 3.2.1: Positioning of point

Now see in Fig. (3.2.1), let ABCD the surface of table and the position of lamp is on point L. The position of L is 3-unit distance from line AC along

3. COORDINATE GEOMETRY

AB and 4-unit distance from AB along to line AC. That is exact position of lamp.

Problem 3.2.2. (Street Plan): A city has two main roads which cross each other at the center of the city. These two roads are along the North-South direction and East-West direction.

All the other streets of the city run parallel to these roads and are 200m apart. There are 5 streets in each direction. Using $1cm = 200m$, draw a model of the city on your note book. Represent the roads/streets by single line.

There are many cross-streets in your model. A particular cross-street is made by two-street, one running in the North-South direction another in the East-West direction. Each cross street is referred to in the following manner: If the 2nd street running in the Nortn-South direction and 5th in the East-West direction meet at some crossing, then we will call this cross-street $(2,5)$. Using this convention, find:

   i How many cross-streets can be referred to as $(4,3)$.

   ii How many cross-streets can be referred to as $(3,4)$.

Solution: To draw model of city, let $1cm = 200m$, now draw the main roads which are North-South direction and other in East-West direction. All another 5 streets draw parallel to these two main streets.

3. COORDINATE GEOMETRY

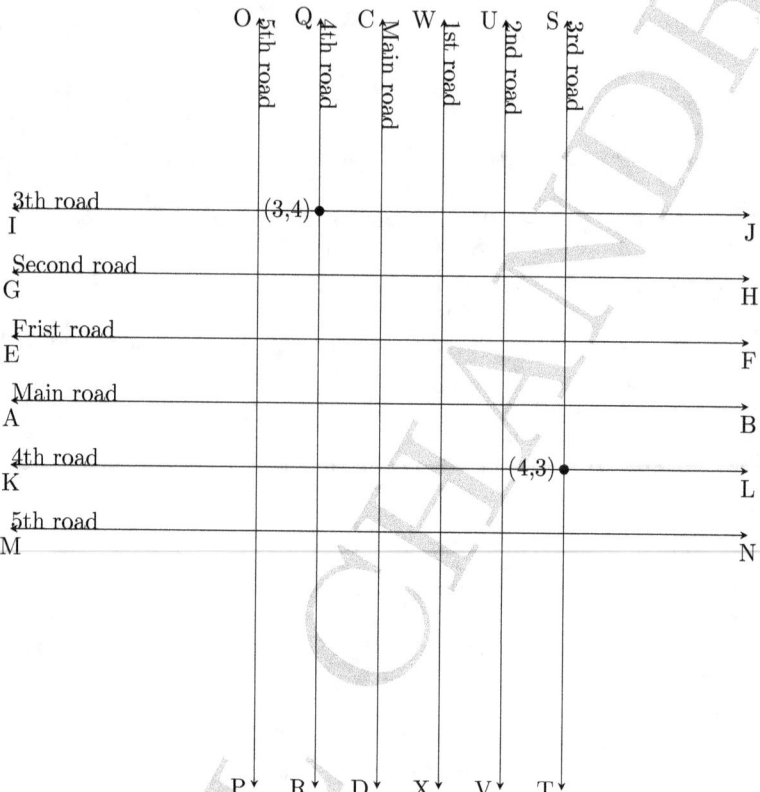

Figure 3.2.2: Model of city

i Only one cross-streets referred to $(4,3)$.

ii Only one cross-streets can be referred to $(3,4)$.

3. COORDINATE GEOMETRY

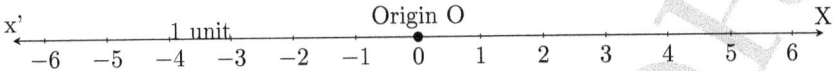

Figure 3.3.1: Ploting of point on horizontal line

3.3 Cartesian System

We have been studied the number line in chapter one. In that chapter we learn that how to place a real number on real line. There are positive and negative real numbers which was correcly placed on real number line. Hence each real number associated to a distance, each distance measured from a fixed point. The fixed point from which distance mesured for a real number called **origin.** The distance measured left side of the origin or right side of the origin represents as negative or positive real numbers. Similar manner You can draw a verticle line as well as horizontao line.

3. COORDINATE GEOMETRY

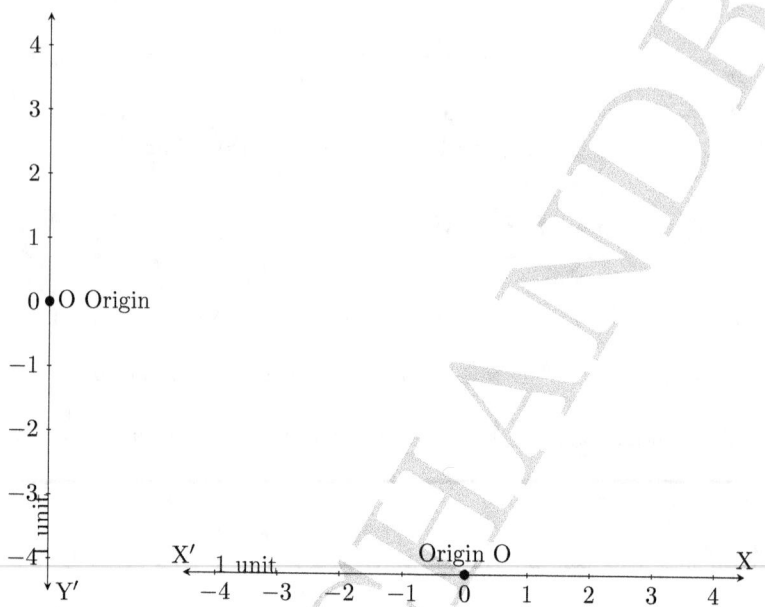

Figure 3.3.2: Ploting of point on vertical and horizontal line

When you combine these two horizontal and vertical line, you find a system that may help you to determine the point position.

Now here we shal develop the system that determine exact position of points in a plane. There needs two lines, one is horizontal and other is verticle line. These lines are called $X'X$ as **x-axis** and $Y'Y$ as **y-axis**. The intersection of x-axis and y-axis is called **origin**. The OX and OY are called the positive direction of $x-$axis and $y-$ axis. Similarly OX' and OY' are called the negative direction of $x-$axis and $y-$ axis. These two lines devide

3. COORDINATE GEOMETRY

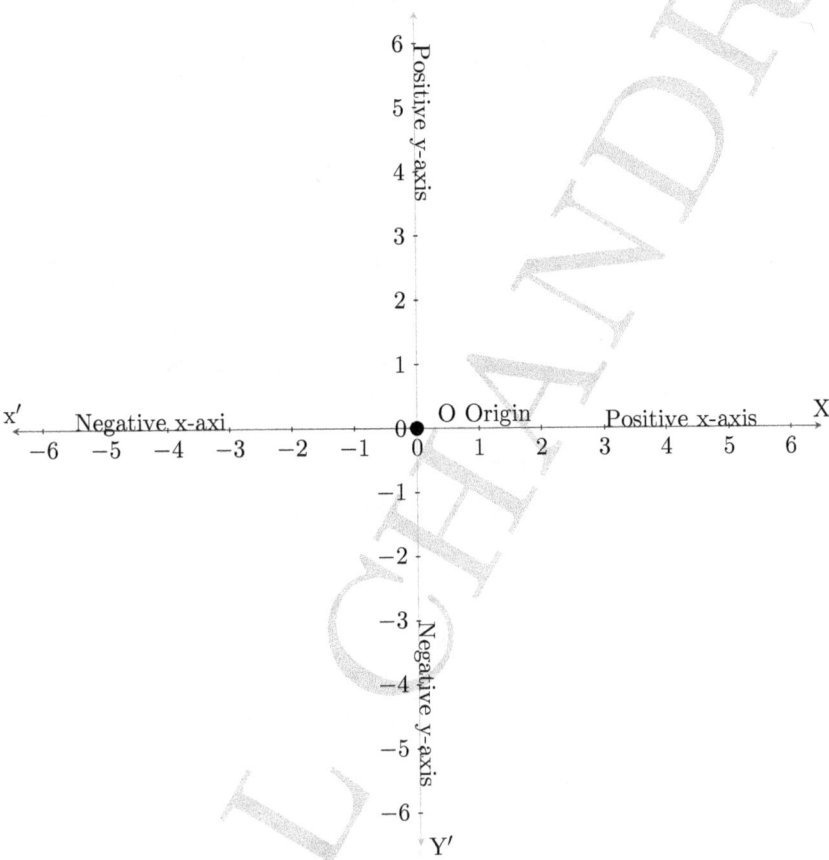

Figure 3.3.3: Coordinate axes

the whole plane in four part, each part called I quadrant, II quadrant, III quadrant, IV quadrant.

To determine a point uniqualy in coordinate plane or $xy-$ plane. There defined two co-ordinate associated to each point in the $xy-$ plane.

3. COORDINATE GEOMETRY

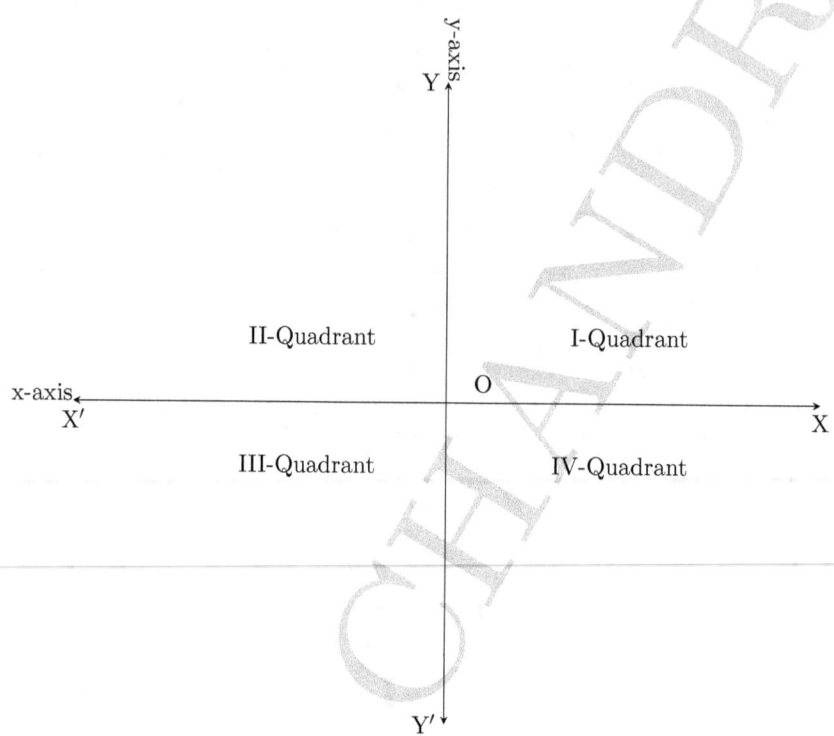

Figure 3.3.4: Coordinate System

Rule to determine the coordinate of a point :

i The $x-coordinate$ of a point is its perpendicular distance from $y-axis$.

The $x - coordinate$ is also called the **abscissa**.

ii The $y-coordinate$ of a point is its perpendicular distance from $x-axis$.

The $y - coordinate$ is also called the **ordinate**.

3. COORDINATE GEOMETRY

iii To writting the coordinate of a point in the coordinate plane, the $x-coordinate$ comes frist, and then the $y-coordinate$. These coordinates place in small brackets.

Example 3.3.1. See the Fig. 3.3.5 and complete the following statemants:

i The abscissa and ordinate of the point B ate ... and ..., respectively. Hence coordinate of B are (...,...).

ii The x-coordinate and y-coordinate of the point M are ... and ..., respectively. Hence coordinate of M are (...,...).

iii The x-coordinate and y-coordinate of the point L are ... and ..., respectively. Hence coordinate of L are (...,...).

iv The x-coordinate and y-coordinate of the point S are ... and ..., respectively. Hence coordinate of S are (...,...).

Solution:

i In Fig.(3.3.5) the distance of point B from the y-axis is 4 units, the x-coordinate or abscissa of the point B is 4. The distance of the point B from x-axis is 2.5 units, the y-coordinate or ordinate of the point B is 2.5. Hence the coordinate of the point B are $(4, 2.5)$.

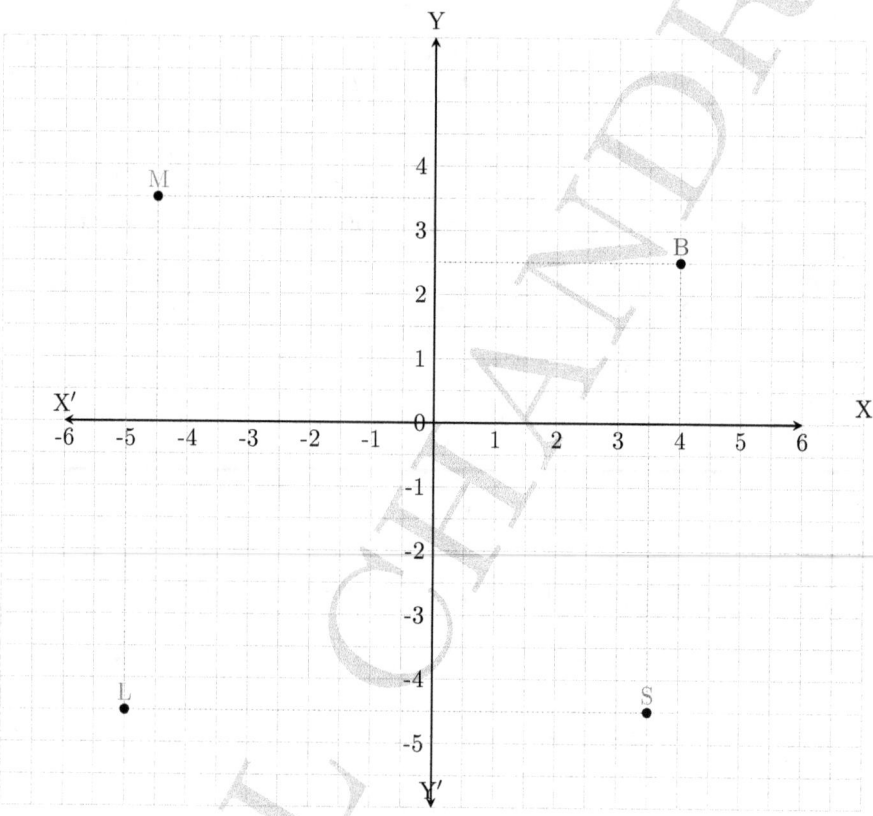

Figure 3.3.5

ii The x-coordinate and y-coordinate of the the point M are -4.5 and 3.5 respectively. Hence, the coordinate of the point M are $(-4.5, 3.5)$.

iii The x-coordinate and y-coordinate of the the point L are -5 and -4.5 respectively. Hence, the coordinate of the point L are $(-5, -4.5)$.

3. COORDINATE GEOMETRY

iv The x-coordinate and y-coordinate of the the point S are 3 and -4.5 respectively. Hence, the coordinate of the point S are $(3, -4.5)$.

Example 3.3.2. Write the coordinate of the points marked on the axes in the Fig.(3.3.6),

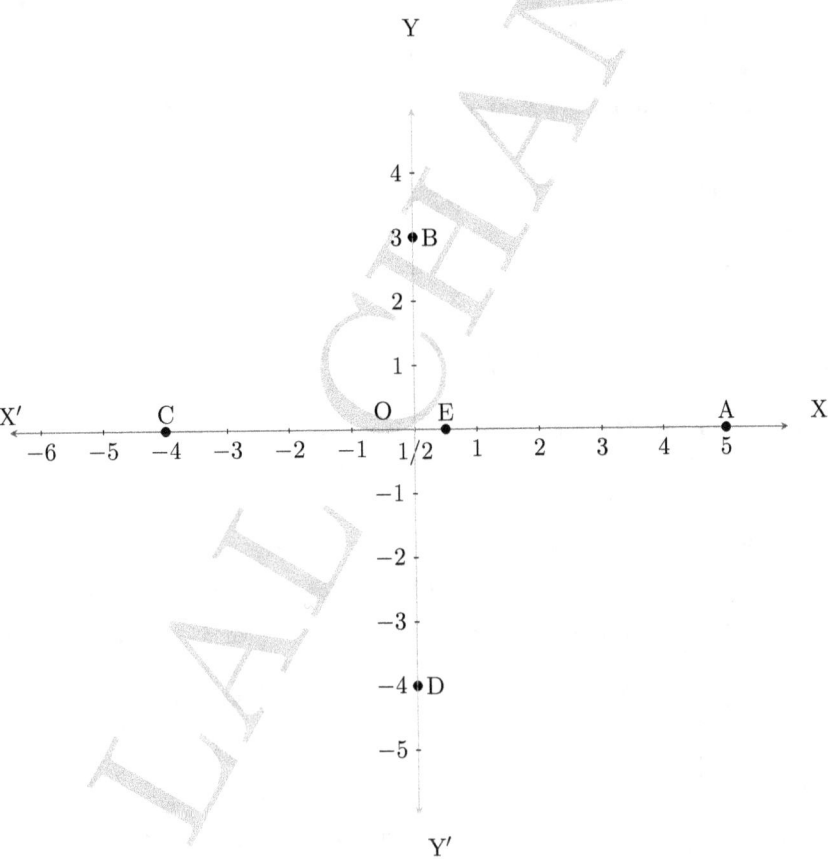

Figure 3.3.6

3. COORDINATE GEOMETRY

Solution: From Fig.(3.3.6),

i The distance of the point A from y-axis is $+5$ and distance from x-axis is 0. Hence the x-coordinate of A is 5 and the y-coordinate is 0. So, the coordinate of A are $(5,0)$. Similar manner the coordinates of following points are given,

ii The coordinate of B are $(0,3)$.

iii The coordinate of C are $(-4,0)$.

iv The coordinate of D are $(0,-4)$.

v The coordinate of E are $(.5,0)$.

Remark: The coordinates of origin are $(0,0)$ because the distance of origin from x-axis and y-axis are 0.

Observation:

i If a point is in the 1st quadrant, then the coordinates of the point will be in the form $(+,+)$, since the 1st quadrant is enclosed by the positive x-axis and positive y-axis.

ii If a point is in the 2nd quadrant, then the coordinates of the point will be in the form $(-,+)$, since the 2nd quadrant is enclosed by the negative x-axis and positive y-axis.

3. COORDINATE GEOMETRY

iii If a point is in the 3rd quadrant, then the coordinates of the point will be in the form $(-,-)$, since the 3rd quadrant is enclosed by the negative x-axis and negative y-axis.

iv If a point is in the 4th quadrant, then the coordinates of the point will be in the form $(+,-)$, since the 4th quadrant is enclosed by the positive x-axis and negative y-axis.

3.4 Solution of Exercise 3.2

Problem 3.4.1. Write the answer of each of the following questions:

i What is the name of horizontal and the vertical lines drawn to determine the position of any point in the Cartesian plane ?

ii What is the name of each part of the plane formed by these two lines ?

iii Write the name of the point where these two lines intersect ?

Solution:

i The horizontal line is called x-axis and vertical line is called y-axis which drawn to determine the position of any point in the cartesien plane.

ii x-axis and y-axis are divide the plane in four patrt, which is called I-quadrant, II-quadrant, III-quadrant, IV-quadrant.

3. COORDINATE GEOMETRY

iii The intersection of x-axis and y-axis is called origin.

Problem 3.4.2. See the Fig.(3.3.7), and write the following:

i The coordinate of B.

ii The coordinate of C.

iii The point identified by the coordinate $(-3, -5)$.

iv The point identified by the coordinate $(2, 4)$.

v The coordinate of D.

vi The coordinate of H.

vii The coordinate of L.

viii The coordinate of M.

Solution:

i The distance of point B from x-axis is 2 in positive direction and y-axis is 5 in negative direction. Hence the coordinates of the point B are $(-5, 2)$.

ii The distance of point C from x-axis is 5 in negative direction and y-axis is 5 negative direction. Hence the coordinates of the point B are $(5, -5)$.

3. COORDINATE GEOMETRY 103

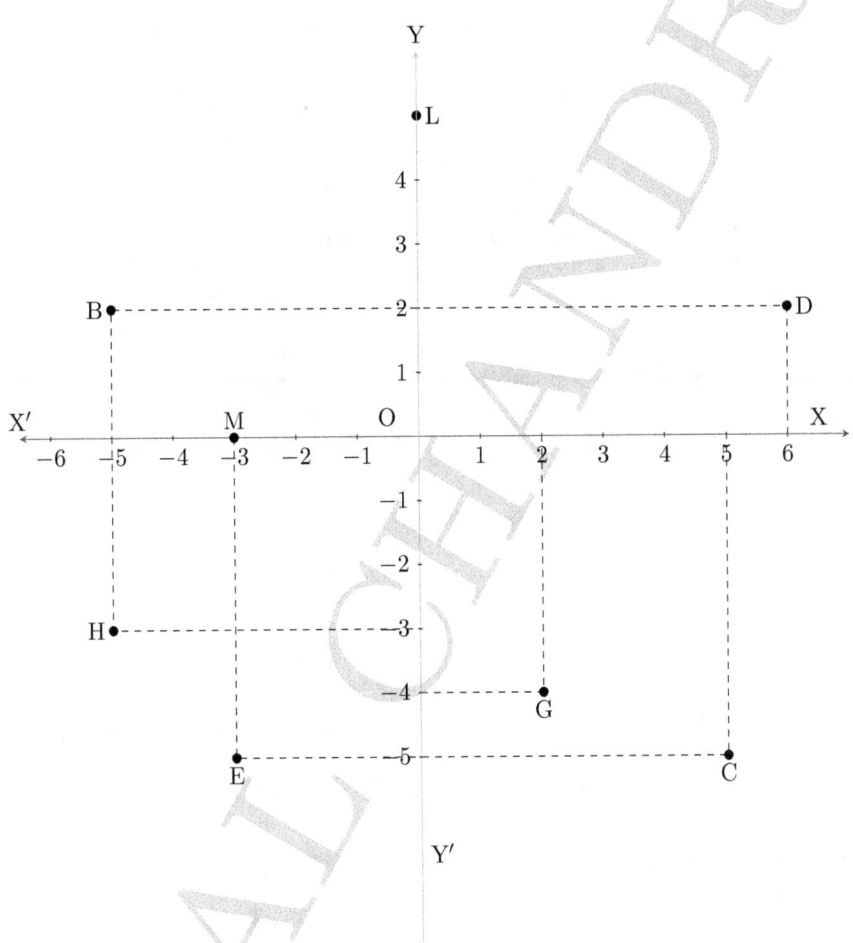

Figure 3.4.1

(iii) The distance of the point E have 5 units in negative direction from x-axis and 3 units in negative direction from y-axis. Hence, the point identified by the coordinates $(-3, -5)$ is E.

3. COORDINATE GEOMETRY

iv The distance of the point E has 4 units in negative direction from x-axis and 2 units in positive direction from y-axis. Hence, the point identified by the coordinates $(2, -4)$ is G.

v The distance of point D from x-axis is 2 in positive direction and y-axis is 6 in negative direction. Hence the coordinates of the point B are $(6, 2)$.

vi The distance of point H from x-axis is 3 in negative direction and y-axis is 5 negative direction. Hence the coordinates of the point B are $(-5, -3)$.

vii The distance of point L from x-axis is 5 in positive direction and y-axis is 0. Hence the coordinates of the point B are $(0, 5)$.

viii The distance of point B from x-axis is 0 and y-axis is 3 in negative direction. Hence the coordinates of the point B are $(-3, 0)$.

4 LINEAR EQUATIONS IN TWO VARIABLES

4.1 Introduction

We have familiar in earlier classes with linear equations in one variable. They are in the form $x + 2 = 0$, $x - \sqrt{5} = 0$, $\pi x + 7 = 0$, $x = 0$. Also we have seen that these equations has unique solutin in real number system. Now there are questions as what is the form of linear equations in two variables, the solution of these equations will be unique or not. Answer of all such type questions will be given in this chapter. The visualization of equatios and its solutions will be consider in this chapter also.

4.2 Linear Equations

Consider the following equations,

$$3x + 2 = 0$$

and

$$2s + 7 = 0$$

the solutions of these two equations are a$x = -\frac{2}{3}$, $x = -\frac{7}{2}$. It can represents on number line uniquely. Some properties of equation; the Solution of equation is not affected when:

4. LINEAR EQUATIONS IN TWO VARIABLES

i the same number is added to (or subtracted from) bothe sides of the equation.

ii you multiply or divide both the side of the equationby the same non-zero number.

The equation in form $ax + by + c = 0$ is standard form of **linear equation in two variables** where a, b and c are real numbers, and a and b are not both zero.

Example 4.2.1. Write each of the following equations in the form $ax+by+c = 0$ and indicate the values of a, b and c in each case:

Solution: In the frist column given equation and second column the equation written in the standed form $ax + by + c = 0$.

equation	standard form	values of a, b, c
(i) $2x + 3y = 4.37$	$2x + 3y - 4.37 = 0$	$a = 2, b = 3, c = -4.37$
(ii) $x - 4 = \sqrt{3}y$	$x - \sqrt{3}y - 4 = 0$	$a = 1, b = \sqrt{3}, c = -4$
(iii) $4 = 5x - 3y$	$5x - 3y - 4 = 0$	$a = 5, b = -3, c = -4$
(iv) $2x = y$	$2x - y = 0$	$a = 2, b = -1, c = 0$

Example 4.2.2. Write each of the following as an equation in two variables:

i $x = -5$

ii $y = 2$

4. LINEAR EQUATIONS IN TWO VARIABLES

iii $2x = 3$

iv $5y = 2$

Solution:

equation	standard form
(i) $x = -5$	$x + 0y + 5 = 0$
(ii) $y = 2$	$0x + y - 2 = 0$
(iii) $2x = 3$	$2x + 0y - 3 = 0$
(iv) $5y = 2$	$0x + 5y - 2 = 0$

4.3 Solution of Exercise 4.1

Problem 4.3.1. The cost of a notebook is twice the cost of a pen. Write a linear equation in two variable to represent this statement.

Solution: Let the cost of a notebook is Rs.x and cost of a pen is Rs. y. Then according to the question,

$$x = 2y.$$

Problem 4.3.2. Express the following linear equations in the form $ax + by + c = 0$ and indicate the values of a, b and c in each case:

i $2x + 3y = 9.3\bar{5}$

ii $x - \frac{y}{5} - 10 = 0$

4. LINEAR EQUATIONS IN TWO VARIABLES

iii $-2x + 3y = 6$

iv $x = 3y$

v $2x = -5y$

vi $3x + 2 = 0$

vii $y - 2 = 0$

viii $5 = 2x$

Solution: The standard form of given equations and there coefficients are:

equation	standard form	values of a, b, c
(i) $2x + 3y = 9.3\bar{5}$	$2x + 3y = 9.3\bar{5}$	$a = 2, b = 3, c = 9.3\bar{5}$
(ii) $x - \frac{y}{5} - 10 = 0$	$x - \frac{y}{5} - 10 = 0$	$a = 1, b = -\frac{1}{5}, c = -10$
(iii) $-2x + 3y = 6$	$-2x + 3y - 6 = 0$	$a = -2, b = 3, c = -6$
(iv) $x = 3y$	$x - 3y + 0 = 0$	$a = 1, b = -3, c = 0$
(v) $2x = -5y$	$2x + 5y + 0 = 0$	$a = 2, b = 5, c = 0$
(vi) $3x + 2 = 0$	$3x + 0y + 2 = 0$	$a = 3, b = 0, c = 2$
(vii) $y - 2 = 0$	$0x + y - 2 = 0$	$a = 0, b = 1, c = -2$
(viii) $5 = 2x$	$2x + 0y - 5 = 0$	$a = 2, b = 0, c = -5$

4. LINEAR EQUATIONS IN TWO VARIABLES

4.4 Solution of a Linear Equation

We have been seen that every linear equation in one variable has unique solution. But what we can say about the solution of linear equation in two variables. Frist there is a question that what are the solution of linear equation in two variable. The values of x and y which satisfy the linear equation in two variable is called the **solution** of equation. **Remark:**

- There are infinite number of solutions to a linear equation in two variables.

Example 4.4.1. Find four solution of the equation $x + 2y = 6$.

Solution: Now by inspection, take $x = 4, y = 1$, satisfy the equation, so $(4, 1)$ is a solution of the equation $x + 2y = 6$. As,

$$x + 2y = 4 + 2 \times 1 = 6$$

Consider $x = 2$ then $2 + 2y = 6 \implies 2y = 4 \implies y = 2$. Hence $(4, 2)$ is a solution. Similar manner if $x = 6$ then $6 + 2y = 6 \implies 2y = 0 \implies y = 0$. Hence $(6, 0)$ is solution. If $x = 0$ then $0 + 2y = 6 \implies y = \frac{6}{2} \implies y =$. Hence $()0, 3$ is the solution of equation.

Example 4.4.2. Finf two solutions for each of the following equations:

i $4x + 3y = 12$

4. LINEAR EQUATIONS IN TWO VARIABLES

ii $2x + 5y = 0$

iii $3y + 4 = 0$

Solution:

i Taking $x = 0$, We get $4 \times 0 + 3y = 12 \implies 3y = 12 \implies y = 4$, so $(0, 4)$ is a solution of equation. If we take $y = 0$, we get $x = 3$ so $(4, 0)$ second solution of equation.

ii Taking $x = 0$, we get $5y = 0 \implies y = 0$. Hence $(0, 0)$ is a solution of given equation. Again take $x = 1$ we get $2 + 5y = 0 \implies y = -\frac{2}{5}$. Hence $\left(1, -\frac{2}{5}\right)$ is a solution.

iii The given equation is $3y + 4 = 0$. In this equation x variable is not present hence x can take any values. From the equation we get $y = -\frac{4}{3}$, therefore $\left(0, -\frac{4}{3}\right), \left(1, -\frac{4}{3}\right)$ are two solution of given equation.

4.5 Solution of Exercise 4.2

Problem 4.5.1. Which one of the following option is true, and why?
$y = 3x + 5$ has

i a unique solution,

ii only two solution,

4. LINEAR EQUATIONS IN TWO VARIABLES

iii infinitely many solutions.

Solution: The equation $y = 3x + 5$ has infinite many solutions so option (iii) is true. Because for each x, equation gies a value of y.

Problem 4.5.2. Write four solutions for each of the following equations:

i $2x + y = 7$

ii $\pi x + y = 9$

iii $x = 4y$

solution:

i Take $x = 0$, we get $y = -2x + 7 \implies y = -2 \times 0 + 7 \implies y = 7$. Hence $(0, 7)$ is a solution. Next take $y = 0$, so $2x + 0 = 7 \implies x = \frac{7}{2}$. Hence $(\frac{7}{2}, 0)$ is another solution. Now take $x = 1$, then we get $2 \times 1 + y = 7 \implies y = 5$ so $(1, 5)$ is a solution also. Next take $x = -1$, then $2(-1) + y = 7 \implies y = 9$. Hence $(-1, 9)$ is a solution.

ii Take $x = 0$, we get $y = -\pi x + 9 \implies y = -\pi \times 0 + 9 \implies y = 9$. Hence $(0, 9)$ is a solution. Next take $y = 0$, so $\pi x + 0 = 9 \implies x = \frac{9}{\pi}$. Hence $(\frac{9}{\pi}, 0)$ is another solution. Now take $x = 1$, then we get $\pi \times 1 + y = 9 \implies y = 9 - \pi$ so $(1, 9 - \pi)$ is a solution also. Next take $x = -1$, then $\pi(-1) + y = 9 \implies y = 9 + \pi$ Hence $(-1, 9 + \pi)$ is a solution.

4. LINEAR EQUATIONS IN TWO VARIABLES

iii Take $x = 0$, we get $4y = x \implies y = 0$. Hence $(0,0)$ is a solution. Next take $y = 1$, so $4 \times 1 = x \implies x = 4$. Hence $(4,1)$ is another solution. Now take $y = 2$, then we get $x = 4 \times 2 \implies x = 8$ so $(8,2)$ is a solution also. Next take $y = -1$, then $x = (-1)4 \implies x = -8$. Hence $(-8,-1)$ is a solution.

Problem 4.5.3. Check which of the following are solutions of the equation $x - 2y = 4$ and which are not:

i $(0,2)$

ii $(2,0)$

iii $(4,0)$

iv $(\sqrt{2}, 4\sqrt{2})$

v $(1,1)$

Solution:

i now putting $x = 0, y = 2$ in the equation $x - 2y = 4$. We get $0 - 2 \times 2 = 4 \implies -4 = 4$, which absurd. So $(0,2)$ is not solution of the equation $x - 2y = 4$.

4. LINEAR EQUATIONS IN TWO VARIABLES

ii now putting $x = 2, y = 0$ in the equation $x - 2y = 4$. We get $2 - 2 \times 0 = 4 \implies 2 = 4$, which absurd. So $(2, 0)$ is not solution of the equation $x - 2y = 4$.

iii now putting $x = 4, y = $ in the equation $x - 2y = 4$. We get $4 - 2 \times 0 = 4 \implies 4 = 4$, So $(0, 2)$ is a solution of the equation $x - 2y = 4$.

iv now putting $x = \sqrt{2}, y = 4\sqrt{2}$ in the equation $x - 2y = 4$. We get $\sqrt{2} - 2 \times 4\sqrt{2} = 4 \implies -7\sqrt{2} = 4$, which absurd. So $(0, 2)$ is not solution of the equation $x - 2y = 4$.

v now putting $x = 1, y = 1$ in the equation $x - 2y = 4$. We get $1 - 2 \times 1 = 4 \implies -1 = 4$, which absurd. So $(1, 1)$ is not solution of the equation $x - 2y = 4$.

Problem 4.5.4. Find the value of k, if $x = 2, y = 1$ is a solution of the equation $2x + 3y = k$.

Solution If $x = 2, y = 1$ is a solution of the equation $2x + 3y = k$. Then value of x, y will satisfy the equation. Now puttinf the value of x, y in $2x + 3y = k$. We get $2x + 3y = k \implies 2 \times 2 + 3 \times 1 = k \implies k = 4 + 3 = 7$. The value of $k = 7$.

5 INTRODUCTION TO EUCLID'S GEOMETRY

5.1 Introduction:

In the mathematics world, 'geometry' word give the image of their development and place of birth. Geometry made from two word 'geo' and 'metriein' which are Greek word it reffer as measure of eart. So we arrived a conclucion that this branch of mathematics development starts in Egipt. Later other part of world gave their contribution to make rich as a displine. The development foot print of geometry in babylonia, china, india apears.

The first proof in geometry given by Greek mathematician Thales as "a circle is bisected by tis diameter.Thales's pupil pythagoras (572 BCE) and his group discovered many geometric properties and developed the theory of geometry.Next mathematician Euclid which was teacher at Alexandria in Egypt organised all the theorems and properties till 300 BCE in his book 'Elements'.

In this chapter, we will study Euclid's approach to geometry and shal try to link it with the present day geometry.

5.2 Euclid's Definitions, Axioms and Postulates

At the time Greek mathematician Euclid study of geometry in abstract form mostly. They consider the solid as solid has shape, size, position and can be moved from one placed to another. Its boundaries are called **surfaces**. They separate one part of the space from another, are said to have no thickness. The boundaries of the surfaces are **curves** or **lines**. These lines end in **points**.

Now in real world any solid object has three extentions which is called a **dimension.** Hence a solid has three dimensions, a surface has two dimension, aline has one dimension and a point sah none. Euclid organise these statments as definitions in his book **Elements**. A few of them are given below:

i A **point** is that which has no part.

ii A **line** is breadthless length.

iii The end of a line are points.

iv A **straight line** is a line which lies evenly with the points on itself.

v A **surface** is that which has length and breadth only.

vi The edge of a surface are lines.

vii A **plane surface** is a surface which lies evenly with the straight line on itself.

5. INTRODUCTION TO EUCLID'S GEOMETRY

If you study carefully these definitions there are many terms used, which need to define them. Next you need to define terms involve in definitions of terms. So this sequence of definition will continue. Hence some geometric terms consider **undefined**. So in geometry, you should take *a point, a line and a plane (in Euclid word a plane surface) as undefined terms*. But these terms can be explain then with help of *'physical models'*.

To develop the geometry and define definition Euclid assumed certain properties, which were not to be proved. But these assumptions are actually *"obvious universal truths"*. He divided them into two types: axioms and postulates. To Geometrical assumption he called **'postulate'** and common notions that not spacialy related to geometry and use throughout mathematics is called **'axioms'**.

There are some Euclid's Axioms:

    i Things which are equal to same thing are equal to another.

    ii If equals are added to equals, the whole are equal.

    iii If equals are subtracted from equal, the remainder are equal.

    iv Things which coincide with one another are equalto one another.

    v The whole is greater than the part.

5. INTRODUCTION TO EUCLID'S GEOMETRY

vi Things which are double of same things are equal to one another.

vii Things which halves of the same things are equal to one another.

Application of Axiom

The concept of application of axiom:

i The Seven Axiom are applicable for 'common notion' to magnitudes of some kind. The Frist common notion could be applied to plane figures as area of plane figure.

ii The magnitude of the same kind can be compared and added but magnitude of different kinds can not be compared. For example, a line cannot be compared to a rectangle, nor can angle be compared to pentagon.

iii The 4th axiom say that if two things are identical, then they are equal. It can be say that, everything equal itself.

iv Axiom 5th give concept of 'greater than'. It means if quantity B is part of quantity A, Then $A > B$

Postulate 1 : *A straight line may be drawn from any one point to any other point.*

This postulate tells us that there is a *unique line joining two distinct points*.

5. INTRODUCTION TO EUCLID'S GEOMETRY

Axiom 5.1 : *Given two distinct points, there is a unique line that passes through them.*

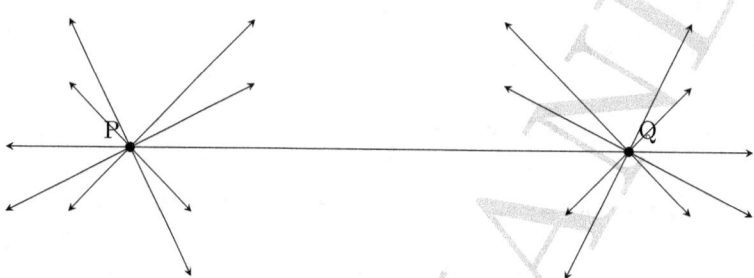

Figure 5.2.1

In the Fig.5.2.1 there are only one line passing through P also pass through Q that is PQ. next there are only one line passing through Q also pass through P that is PQ. Hence Axiom 5.1 is self-evident.

Postulate 2 : *A terminated line can be produced indefinitely.*

This postulate say that any line segment can be extended on either side to form a line.

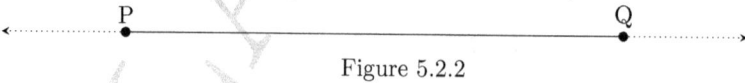

Figure 5.2.2

Postulate 3 : *A circle can be drawn with any centre and any radius.*

Postulate 4 : *All right angle are equal to one another.*

5. INTRODUCTION TO EUCLID'S GEOMETRY

Postulate 5 : *If a straight line falling on two straight lines makes the interior angles on the same side of it taken together less than two right angles, then the two straight lines, if produced infinitely, meet on that side on which the sum of angle is less than two right angles.*

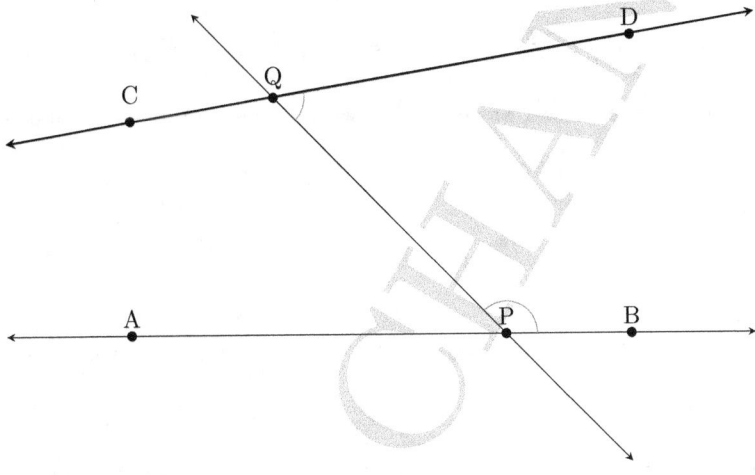

Figure 5.2.3

For example line PQ in the Fig.5.2.3 falls on line AB and CD such that the interior angle Q and P is greater than 180^0 on the right of PQ. Therefore, the lines AB and CD will not intersect on left right side of PQ.

On the basis of his postulate and axiom Euclid prooved other results. Then, the basis of these results he proved more results applying deductive reasoning. These proved statements are called **propositions or theorems**.

Example 5.2.1. If A, B and C are three points on a line, and B lies between

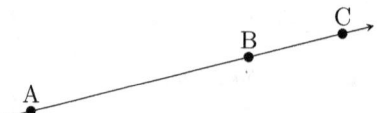

Figure 5.2.4

A ans C (With Fig.5.2.4), then prove that $AB + BC = AC$.

Solution: In the Fig.5.2.4 AC coincide with AB+BC. Hence by Euclid Axiom (4), tells that things which coincide with one another are equal to one another. So, it can be deducted that

$$AB + BC = AC.$$

Example 5.2.2. Prove that an equilateral tringle can be constructed on any given line segment.

Solution: In the given statement, a line segment of any length is given, Say AB, in (Fig.5.2.5).

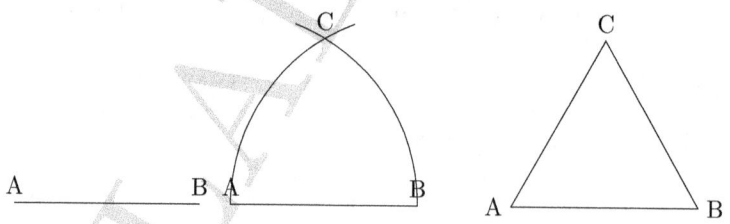

Figure 5.2.5

Here we construct a line segment AB. From Euclid's postulate 3, we can draw a circle with radius AB and center A. Similarly, we draw another ciecle

5. INTRODUCTION TO EUCLID'S GEOMETRY

with radius BA and center B. These two circles meet at a points, say C. Next drawing the line segment AC and BC to form △ ABC.

Now we prove that this triangle is equilateral triangle, i.e, AB=AC=BC.

(1) Here, AB=AC, since they are the radii of same circle.

(2) Similarly, AB=BC, raddi of same circle.

From these two fact and Euclid's axiom (3) that things which are equal to same thing are equal to another, we conclude that AB=BC=AC. So △ ABC is an equilateral triangle.

Theorem 5.2.3: *Two distinct lines connot have more than one point in common.*

Proof. At the same time, let us suppose that the two lines intersec in two distinct points, say P and Q. So, we find two lines passing thruogh two distinct points P and Q. But our assumption clashes with the axiom (5.1) that only one line can pass through two distinct points. Hence, the assumption that we started with, that two lines can pass through through two distinct points is worng. From this, we arrived conclusion that two distinct lines cannot have more than one point in common. □

5. INTRODUCTION TO EUCLID'S GEOMETRY

5.3 Solution of exercise 5.1

Problem 5.3.1. Which of the following statements are true and which are false ? Give reasons for your answer:

i Only one line can pass through a single point.

ii There are an infinite number of lines which pass through two distinct points.

iii A terminated line can be produced indefinitely on both the sides.

iv If two circle are equal, then their radii are equal.

v In Fig.5.3.1, If AB=PQ and PQ=XY, then AB=XY.

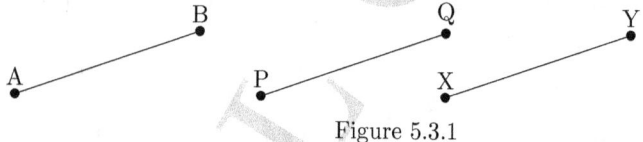

Figure 5.3.1

Solution:

i False. The postulate (1) tells that any draw from one point to one another point. So if you fixed a point, then there are infinite number points around its.

5. INTRODUCTION TO EUCLID'S GEOMETRY

ii False. According axiom (5.1) only one line can be drw from two distinct points.

iii True. According postulate (2) a trminated line can be extended on either side to form a line.

iv True. According axiom (4) if two things are identical, then they are equal. So their radii will be equal.

v True. In the given Fig.5.3.1 given that AB=PQ and PQ =XY. So from Axiom (1), AB=XY.

Problem 5.3.2. Give a definition for each of the following terms. Are there other terms that needed to be define first ?What are they, and how might you define them ?

i Parallel lines.

ii Perpendicular lines.

iii Line segment.

iv Radius of a circle.

v Square.

5. INTRODUCTION TO EUCLID'S GEOMETRY

Solution: To explain the following terms we need define a quantity 'angle'. The angle between to lines is difference of slope these two lines from given a fixed line.

  i Two lines are called parallel line if angle between them is zero degree.

  ii Two lines are called perpendicular if angle between them is 90^0 or a right angle.

  iii A line segment is a part of line which two terminating ends are points.

  iv The distance from center to any point on circumference of circle is called radius of circle.

  v A quadrilateral whose each side have same length and each angles are rightangle.

Problem 5.3.3. Consider two 'postulates' given below:

  i Given any two distinct points A and B, there exists a third point C which is in between A and B.

  ii There exist at least three points that are not on the same line.

Do these postulates contain any undefined terms ? Are these postulates consistent ? Do they follow from Euclid'postulates ? Explain.

5. INTRODUCTION TO EUCLID'S GEOMETRY

Solution: Yes, these postulates contain one undefine terms which is 'distinct point'. Yes, these postulates consistent. Yes, they follow from Eucid's postulates.

Explanation: How to define distinct point because according to Euclid's definition, a point is that which has no part. These postulates does not clashes any postulates given by Euclid. These postulate contained in the Axiom (5.1).

Problem 5.3.4. If a point C lies between two points A and B such that AC=BC, then prove that $AC = \frac{1}{2}AB$. Explain by drawing the figure.

Proof. In the Fig.5.3.2, given that AC=BC. So, 2AC=2BC from axiom (vi). Now AB= 2AC, and AB = 2BC, so AB =2AC=2BC next $\frac{1}{2}AB = AC$ and $\frac{1}{2}AB = BC$ from axiom(vii) $\frac{1}{2}AB = AC$.

A———————C———————B

Figure 5.3.2

□

Problem 5.3.5. In Problem (5.3.4) point C is called a mid point of line segment AB. Prove that every line segment has one and only one mid-point.

Proof. In Fig.5.3.3, suppose, two point C and D are mid-point of line segment AB. From result problem (5.3.4) we got $\frac{1}{2}AB = AC$ and $\frac{1}{2}AB = AD$ because

5. INTRODUCTION TO EUCLID'S GEOMETRY

C and D bothe mid-point of line AB. So Euclid's axiom (i), AD=AC. Hence D and C are identical points. After considaration of these fact we arrived the conclusion that every line segment has one and only one mid-point.

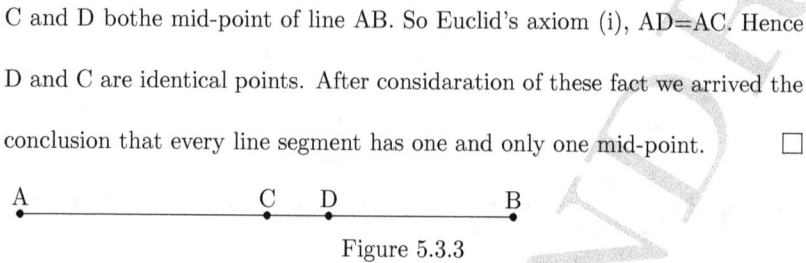

Figure 5.3.3

Problem 5.3.6. In Fig.(5.3.4), if AC=BD, then prove that AB=CD.

Figure 5.3.4

Proof. In Fig.(5.3.4), given that AC = BD. Next we subtract BC to bothe side in equation AC = BD. So AC-BC=BD-BC. Hence with axiom (iii), AB=CD.

Problem 5.3.7. Why is axiom(v), in the list of Euclid's axioms, considered a 'universal truth' ? (Note that the question is not about the fifth postulate.)

Solution: The axiom (v), is true for any quantity of universe, so it is called 'universal truth'.

6 LINES AND ANGLES

6.1 Introduction

In the previous chapter 5, you have learned about points, and lines. You have also learned some axioms and you proved many statements with help of axioms.In this chapter you will study the properties of the angles formed when two lines intersect each other, and also the properties of the angles when two or more parallel lines intersected by a line at two distinct points. Using deductive reasoning you will prove some statements using these propertis.

This knowlwdge needs in your daly life uses and sciensce study. In next chapter of geometry, you will need these propertis of lines and angles to deduce more and more useful properties and statements.

6.2 Basic Terms and Definitions and Notations

Line-segment and Rays: A part (or portion) of a line with two end points is called a **line-segment** and a part of a line with one end point is called a **ray**.

Remark 6.2.1. The line-segment AB is denoted by $\bar{AB}$, and its length is denoted by AB. The ray denoted by $\vec{AB}$, and a line denoted by $\overline{AB}$. However,

6. LINES AND ANGLES

we will not use these symbols, and will denote the line segment AB, ray AB, length AB and line AN by same symbol, AB.

Collinear points and Non-collinear points: If three or more points lie on the same line, they are called **collinear points**, otherwise they are called **non-collinear points**.

Angles and Arms and Vertex: When two rays orginated from same end point, they formed **angle**. These rays which forms angle is called **arms** of angle and the end point is called **vertex** of the angle.

Types of Angles: An **acute angle** measures between 0° and 180°. A **right angle** exactely equal to 90°.

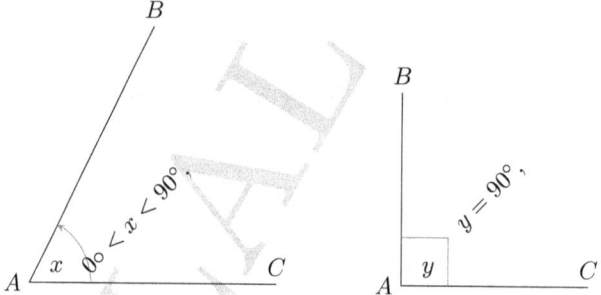

Figure 6.2.1: Acute angle and Right angle

An angle graeter than 90° but less than 180° is called an **obtuse angle**. A **straight angle** equal to 180°.

6. LINES AND ANGLES

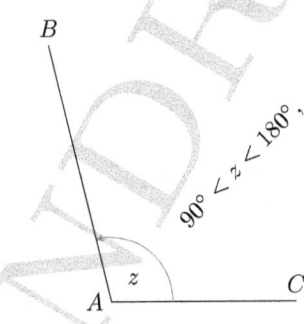

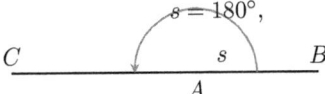

Figure 6.2.2

An angle which is greater than $180°$ but less than $360°$ is called a **reflex angle**.

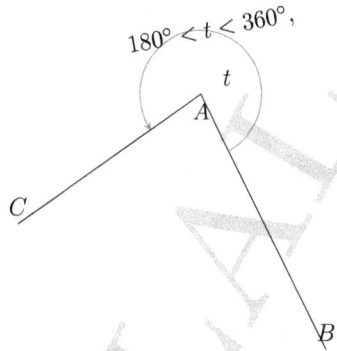

Figure 6.2.3

Two angles are **adjacent**, if they have a common vertex, a common arm and their non-common arms are on different sides of the common arm (See

Fig.6.2.2).

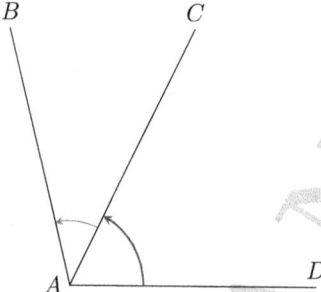

Figure 6.2.4: Adjacent angles

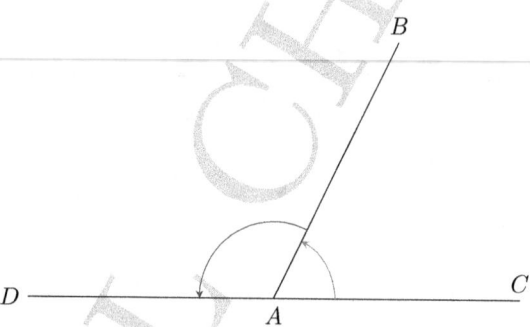

Figure 6.2.5: Linear pair angles

The two angle whose sum is 90° are called **complementry angles**, two angle whose sum is 180° are called **supplementary angles**.

Vertically opposite angle: When two line intersect each other they form vertically opposite angle.

6. LINES AND ANGLES

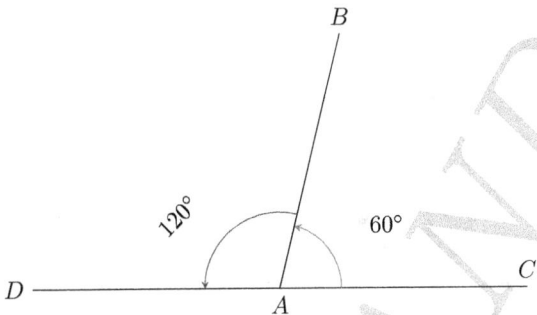

Figure 6.2.6: Supplementary angles

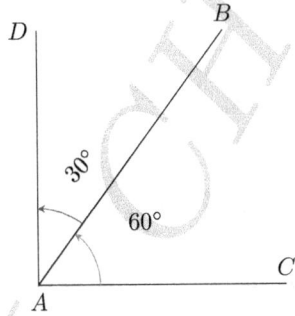

Figure 6.2.7: Complentary angles

Axiom 6.1: If a ray stands on a line, then the sum of two adjacent angle so formed is 180°.

Note : If the sum of two adjacent angle 180°, then a ray stands on line.

Axiom 6.2 : If the sum of two adjacent angle is 180°, then the non-common arms of angles form a line.

The axiom (6.1) and (6.1) together is called the **Linear Pair Axiom**.

6. LINES AND ANGLES

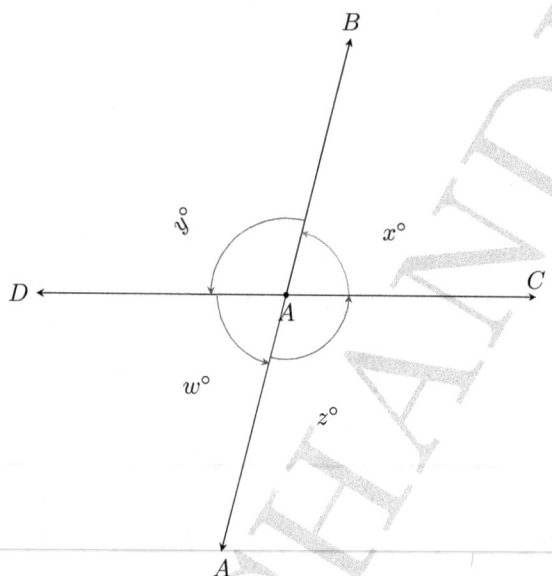

Figure 6.2.8: Vertically opposite angles

Theorem 6.2.2: *If two line intersect each other, then the vertically opposite angles are equal.*

Proof. In the theorem (6.2.2) it is given that 'two lines intersect each other'.

So, let PQ and RS be two lines intersecting at O as given in Fig.(6.2.9).

They lead to two pairs of vertically opposite angles, namely,

(i) $\angle POS$ and $\angle ROQ$ (ii) $\angle ROP$ and $\angle SOQ$.

We need to prove that $\angle POS = \angle ROQ$ and $\angle ROP = \angle SOQ$.

Now ray OP stands on line RS. Therefore,

(6.2.1) $$\angle POS + \angle ROP = 180°$$

6. LINES AND ANGLES

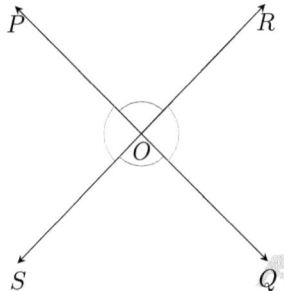

Figure 6.2.9: Vertically opposite angles

(Linear pais axiom)

(6.2.2) $$\angle SOR + \angle QOR = 180°$$

From equation (6.2.1) and (6.2.2) we get

$$\angle POS + \angle ROP = \angle ROP + \angle QOR$$

Thisimplies $\angle POS = \angle QOR$

Similarly, it can be proved that $\angle POR = \angle QOS$. □

Example 6.2.3. In the Fig. (6.2.10) Line PQ and RS intersect each other at point O. If $\angle POR : \angle ROQ = 5 : 7$, find all the angles.

Solution : From theorem (6.2.2), we have $\angle POR : \angle ROQ = 180°$ as Linear pair of angle.

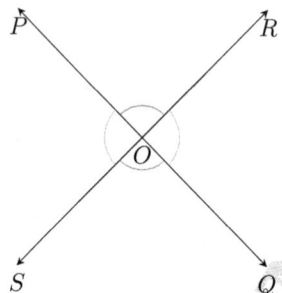

Figure 6.2.10

But given that $\angle POR : \angle ROQ = 5 : 7$.

Therefore,

$$\angle POR = \frac{5}{12} \times 180° = 75°$$

Similarly,

$$\angle ROQ = \frac{7}{12} \times 180° = 105°$$

Now, $\angle POR = \angle SOQ = 75°$ because they are vertically opposite angle.

$\angle ROQ = \angle POS = 105°$ because they are vertically opposite angle.

Example 6.2.4. In Fig. (6.2.11) ray OS stands on a line POQ. Ray OR and ray OT atr angle bisectors of $\angle POS$ and $\angle SOQ$, respectively. If $\angle POS = x$, find $\angle ROT$.

Solution : Ray OS stands on a line POQ. Therefore,

$$\angle SOQ + \angle SOP = 180°$$

6. LINES AND ANGLES

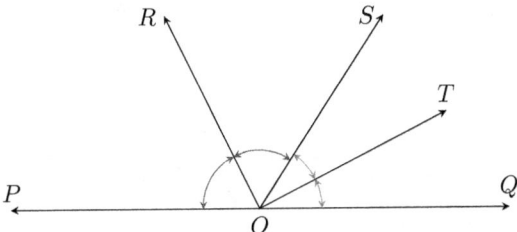

Figure 6.2.11

Given, $\angle POS = x$. Therefore,

$$x + \angle SOQ = 180°$$

So,

$$\angle SOQ = 180° - x$$

Now, ray OR bisects $\angle POS$, therefore,

$$\angle ROS = \frac{1}{2} \times \angle POS = \frac{1}{2} \times x = \frac{x}{2}$$

Similaraly,

$$\angle SOT = \frac{1}{2} \times \angle SOQ = \frac{1}{2} \times (180 - x) = 90° - \frac{x}{2}$$

Now,

$$\angle ROT = \angle ROS + \angle SOT \implies \angle ROT = \frac{x}{2} + 90° - \frac{x}{2}$$

$$\angle ROT = 90°$$

6. LINES AND ANGLES

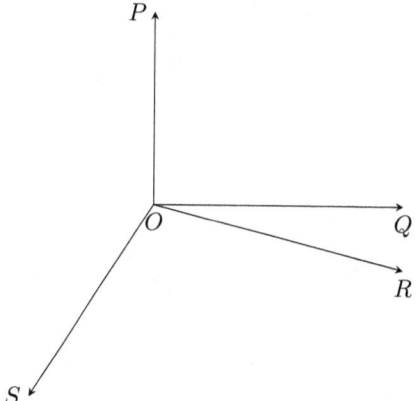

Figure 6.2.12

Example 6.2.5. In Fig.6.2.12 OP, OQ, OR, and OS are four rays. Prove that $\angle POQ + \angle QOR + \angle SOR + \angle POS = 360°$

Solution :

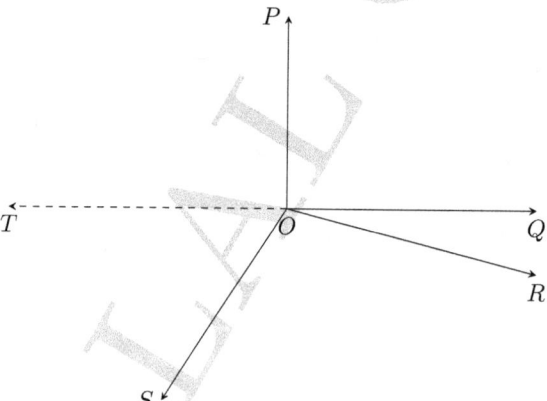

Figure 6.2.13

6. LINES AND ANGLES

In Fig.6.2.12, you need to produced any of the rays OA, OP, OS, OR backwards a point. Let us produce aray OQ backwards to a point T, so TOQ forms a line. See the Fig.(6.2.13).

Now, rays OS stands on line TOQ. Therefore

(6.2.3) $$\angle TOS + \angle SOQ = 180°$$

But $\angle SOQ = \angle SOR + \angle QOR$, so (6.2.3) becomes

(6.2.4) $$\angle TOS + \angle SOR + \angle QOR = 180°$$

Next, rays OS stands on line TOQ. Therefore

(6.2.5) $$\angle TOP + \angle QOP = 180°$$

Now adding equation (6.2.4) and (6.2.5),

(6.2.6) $$\angle TOP + \angle QOP + \angle TOS + \angle SOR + \angle QOR = 360°$$

But, $\angle SOR + \angle QOR = \angle SOQ$. THerefore equation (6.2.6) gives, $\angle TOP + \angle QOP + \angle TOS + \angle SOQ = 360°$

6.3 Solution of Exercise 6.1

Problem 6.3.1. In Fig.(6.3.1), line XY and MN intersect at O. If $\angle POY = 90°$ and $a : b = 2 : 3$, find c.

6. LINES AND ANGLES

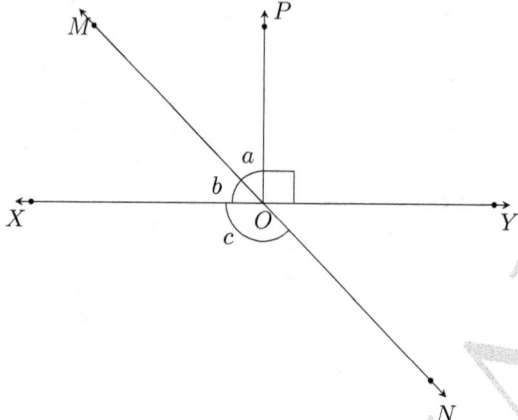

Figure 6.3.1

Solution : In Fig(6.3.1), rays OP stands on line XOY and $\angle POY = 90°$.

Therefore,
$$\angle POY + \angle POX = 180°$$

But, $\angle POY = 90°$, So
$$\angle POY + 90° = 180° \implies \angle POY = 180° - 90° = 90°$$

Now, $\angle MOX = \frac{2}{5} \times \angle POX \implies \angle MOX = \frac{2}{5} \times 90° = 36°$

The ray OX stands on line MON. So,
$$\angle MOX + \angle XON = 180° \implies c + b = 180° \implies c = 180° - b$$

. Therefore, $c = 180° - 36° = 144°$

6. LINES AND ANGLES

Problem 6.3.2. In Fig.(6.3.2), lines AB and CD intersect at O. If $\angle AOC + \angle BOE = 70°$ and $\angle BOD = 40°$, find $\angle BOE$ and reflex $\angle COE$.

Solution : In Fig.(6.3.2), ray OC stans on the line AOB. Therefore,

$$\angle AOC + \angle COB = 180°$$

$$\angle AOC + \angle AOC + \angle BOE = 180°$$

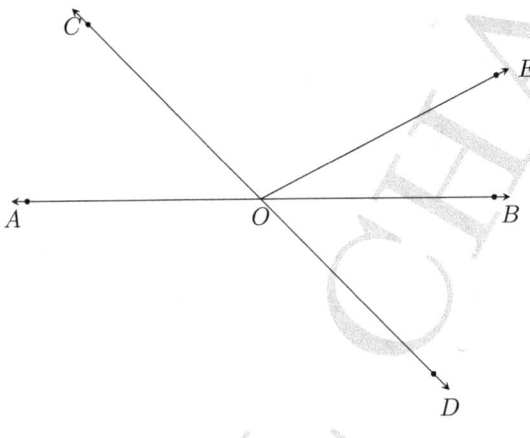

Figure 6.3.2

but, $\angle AOC + \angle BOE = 70°$, so

$$\angle AOC = 180° - 70° = 110°$$

, Hence reflex $\angle ACO = 360° - 110° = 250°$ The lines AB and CD intersec at the point O. So, $\angle AOC = \angle BOD$.

But given that $\angle BOD = 40°$. Therefore, $\angle AOC = 40°$ Now $\angle BOE = 70° - 40° = 30°$.

Problem 6.3.3. In Fig.(6.3.3), $\angle PQR = \angle PRQ$, then prove that $\angle PQS = \angle PRT$.

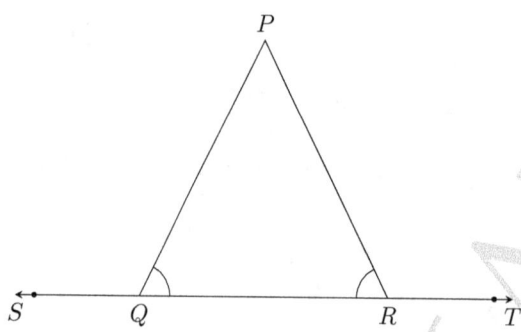

Figure 6.3.3

Solution : In Fig.(6.3.3), The line segments QP and RP stand on line SQRT. Therefore, $\angle SQP, \angle PQR$ and $\angle PRQ, \angle PRT$ are linear pair angle. Hence

$$\angle SQP + \angle PQR = 180°$$

$$\angle PRQ + \angle PRT = 180°$$

Therefore,

$$\angle SQP + \angle PQR = \angle PRQ + \angle PRT$$

But, given that, $\angle PQR = \angle PRQ$. So, $\angle PQS = \angle PRT$.

Problem 6.3.4. In Fig.(6.3.4), if $x + y = w + z$, then prove that AOB is a line.

6. LINES AND ANGLES

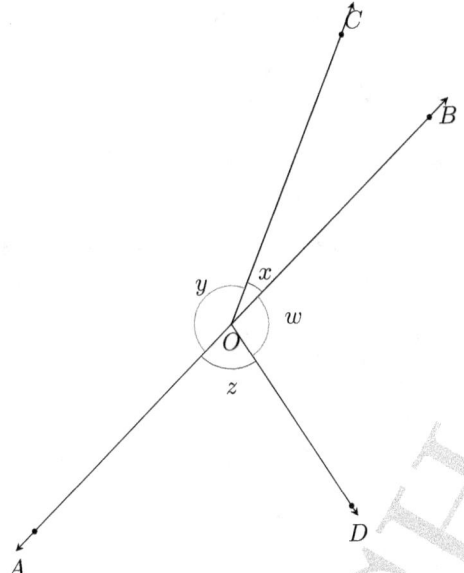

Figure 6.3.4

Solution : In Fig.(6.3.4), rays OA, OB, OC and OD are start from point O. They makes angle x, y, z, w one another. So $x + y + z + w = 360°$, but given that $x + y = w + z$, so

$$2x + 2y = 360°$$

$$x + y = z + w = 180°$$

This gives that

$$\angle BOC + \angle COA = 180°$$

$$\angle AOD + \angle DOB = 180°$$

6. LINES AND ANGLES

Which are adjacent angle, so by axiom (6.2) AOB is a line.

Problem 6.3.5. In Fig.(6.3.5), POQ is a line. Ray OR is perpendicular to line PQ. OS is another ray lying between rays OP and OR. Prove that

$$\angle ROS = \frac{1}{2}(\angle QOS \angle POS).$$

Solution : In Fig.(6.3.5), ray OR stands on line POQ. So,

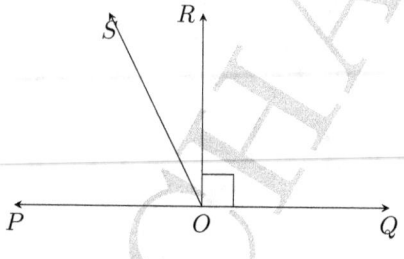

Figure 6.3.5

$$\angle QOR + \angle ROP = 180°$$

(6.3.1) $$\angle QOP = \angle ROP = 90°$$

because ray OR $\perp$ line POQ. From Fig. (6.3.5),

(6.3.2) $$\angle ROS = \angle SOQ - \angle ROQ$$

(6.3.3) $$\angle ROS = \angle POR - \angle POS$$

6. LINES AND ANGLES

From equation (6.3.1), (6.3.2) and (6.3.3),

$$\angle ROS = \frac{1}{2}(\angle QOS \angle POS).$$

Problem 6.3.6. it is given that $\angle XYZ = 64°$ and XY is produced to point P. Draw a figure from the given information. If YQ bisect $\angle ZYP$, find $\angle XYQ$ and refex $\angle QYP$.

Solution : In Fig.(6.3.6), ray YZ stands on line XYP. So,

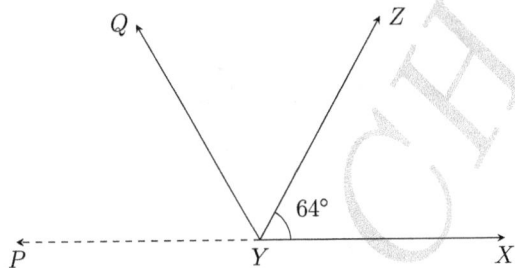

Figure 6.3.6

$$\angle ZYP + \angle ZYX = 180°$$

We have $\angle ZYX = 64°$,

$$\angle ZYP = 180° - 64° = 116°$$

$$\angle XYQ = \frac{1}{2} \times \angle ZYP + \angle ZYX - \frac{1}{2} \times 116° + 64° = 122°$$

Now $\angle QYP = \frac{1}{2} \times 116° = 58°$, so reflex

$$\angle QYP = 360° - 58° = 302°$$

6. LINES AND ANGLES   144

6.4 Lines Parallel to the Same Line

If two lines are parallel to the same line, then they are parallel to each other.

Theorem 6.4.1: *Lines which are parallel to the same line are parallel to each other.*

This theorem can extended more than two lines.

Example 6.4.2. In Fig.(6.4.1), if $PQ \parallel RS$, $\angle MXQ = 135°$ and $\angle MYR = 40°$, find $\angle XMY$.

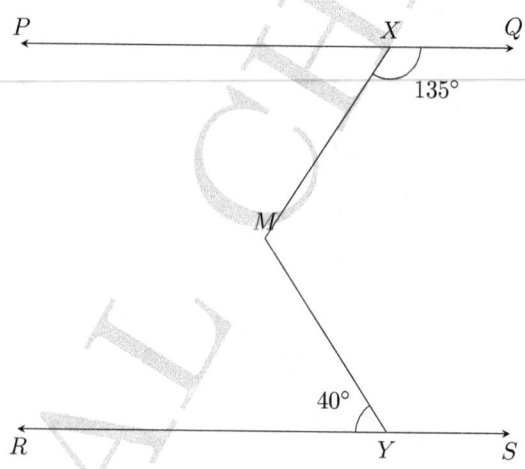

Figure 6.4.1

Solution : In the Fig. (6.4.1), $PQ \parallel RS$, $\angle MXQ = 135°$ and $\angle MYR = 40°$. To find, $\angle XMY$. You construct a line AMB which is parallel to line PQ. Now in Fig. (6.4.2), $PQ \parallel AB$ and $PQ \parallel RS$. Therefore, $AB \parallel RS$. Hence,

6. LINES AND ANGLES

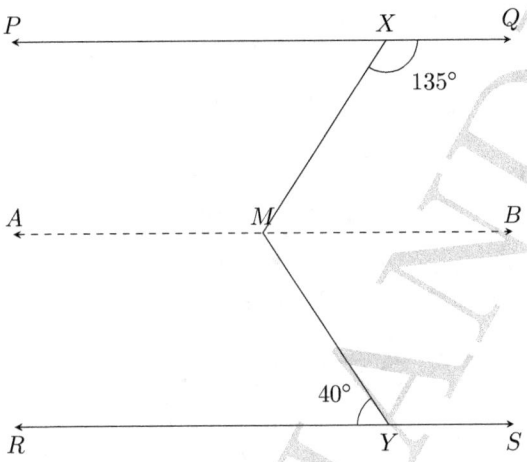

Figure 6.4.2

$$\angle QXM + \angle XMB = 180°$$

($AB \parallel PQ$, interior angles on the same side of the transversal XM)

But
$$\angle QXM = 135°$$

So,
$$\angle XMB = 180° - 135° = 45°$$

(6.4.1)
$$\angle XMB = 45°$$

Now,
$$\angle BMY = \angle MYR$$

6. LINES AND ANGLES

$(AB \parallel RS$, alternate angles) therefore,

$$\angle BMY = 40°$$

(6.4.2) $$\angle BMY = 40°$$

From equation (6.4.1) and (6.4.2), you get

$$\angle XMB + \angle BMY = 45° + 40°$$

That is,

$$\angle XMY = 85°$$

Example 6.4.3. If the transversal intersects two lines such that the bisectors of a pair of corresponding angles are parallel, then prove that the two lines are parallel.

Solution : In Fig. (6.4.3), a transversal AD intersects two lines PQ and RS at points B and C respectively. Ray BE is the bisector of $\angle QBA$ and ray CG is the bisector of $\angle SCB$; $CG \parallel BE$.

We are to prove that $PQ \parallel RS$.

it is given that ray BE is bisector of $\angle ABQ$

Therefore,

(6.4.3) $$\angle EBA = \frac{1}{2}\angle QBA$$

6. LINES AND ANGLES

Similaraly, ray CG is the bisector of ∠SCB

Therefore,

(6.4.4) $$\angle GCB = \frac{1}{2}\angle SCB$$

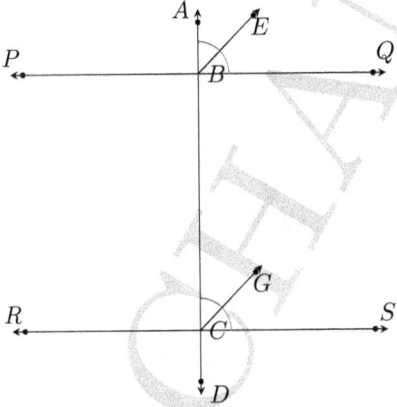

Figure 6.4.3

But, $CG \parallel BE$ and AD is transversal. Therefore,

(6.4.5) $$\angle EBA = \angle GCB$$

(Corresponding angle axiom)

From Equation (6.4.3) and (6.4.4) with (6.4.5), we get

$$\frac{1}{2}\angle QBA = \frac{1}{2}\angle SCB$$

$$\angle SCB = \angle QBA$$

6. LINES AND ANGLES

But the $\angle SCB$ and $\angle QBA$ are corresponding angles formed by transversal line AD with PQ and RS; and equal.

Therefore,

$$RS \parallel PQ$$

(By converse of corresponding angle axiom)

Example 6.4.4. In Fig. (6.4.4), $AB \parallel CD$ and $CD \parallel EF$. Also $EA \perp AB$. If $\angle BEF = 55°$. find the value of x y and z.

Solution : In Fig. (6.4.4), In Fig. (6.4.4), $AB \parallel CD$ and $CD \parallel EF$. Also $EA \perp AB$. If $\angle BEF = 55°$.

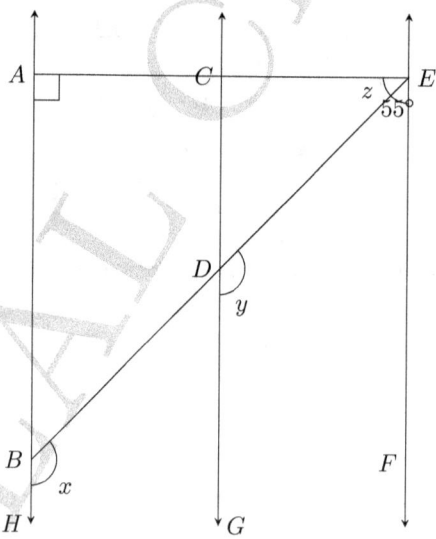

Figure 6.4.4

6. LINES AND ANGLES

The transversal line BE form interior angle $\angle DEF$ and $\angle GDE$ at same side. Therefore,
$$y + 55° = 180°$$
$$y = 180° - 55° = 125°$$

Next,
$$x = y$$
because they are corresponding angle. Therefore,
$$x = 125°$$

So,
$$\angle AED + \angle DEF = 90°$$
$$z = 90° - 55° = 35°$$

6.5 Solution of Exercise 6.2

Problem 6.5.1. In Fig. (6.5.1), if $AB \parallel CD$, $CD \parallel EF$ and $y : z = 3 : 7$, find x

Solution : In Fig. (6.5.1), the line $AB \parallel CD$, $CD \parallel EF$ and $y : z = 3 : 7$. The transversal line PR intresect line AB, CD and EF.

So,
$$\angle DQP = \angle FRQ$$

6. LINES AND ANGLES

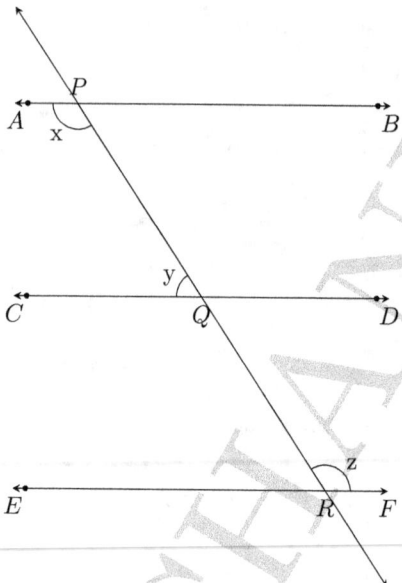

Figure 6.5.1

$$\angle DQP = z$$

This give, $y + z = 180°$ (linear pair angle)

So,
$$y = \frac{3}{10} \times 180° = 54°$$

$$y = \frac{7}{10} \times 180° = 126°$$

Now $x = z = 126°$ (internal alternate angle)

Problem 6.5.2. In Fig. (6.5.2), if $AB \parallel CD$, $EF \perp CD$ and $\angle GED = 126°$, find $\angle AGE$, $\angle GEF$ and $\angle FGE$.

6. LINES AND ANGLES

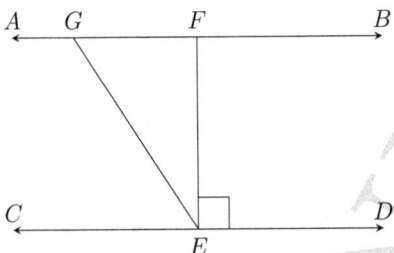

Figure 6.5.2

Solution : In Fig. (6.5.2), the line $AB \parallel CD$, and $EF \perp CD$. So, $EF \perp AB$ and $\angle GFE = 90°$. Given that, $\angle GED = 126°$. In the figure

$$\angle GEF + \angle FED = \angle GED = 126°$$

$$\angle GEF = 126° - 90° = 36°$$

In $\triangle GEF$,

$$\angle EGF = 180° - 36° - 90° = 54°$$

$$\angle AGE = \angle GED = 126°$$

(Corresponding angle)

Problem 6.5.3. In Fig. (6.5.3), if $PQ \parallel ST$, $\angle PQR = 110°$ and $\angle RST = 130°$, find $\angle QRS$.

Solution : Draw a line parallel to ST tsrough point R. In the Fig. (6.5.3). Now in Fig. (6.5.4), $PQ \parallel ST$, $AB \parallel ST$ and $\angle PQR = 110°$ and $\angle RST =$

6. LINES AND ANGLES

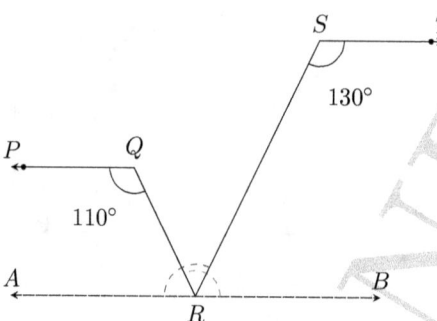

Figure 6.5.3

130°

So,

$$\angle QRB = \angle PQR$$

(Alternate angle axiom)

$$\angle SRA = \angle RSA$$

(Alternate angle axiom)

In the Fig. (6.5.3),

$$\angle ARS + \angle QRB = 180° + \angle QRS$$

$$\angle QRS = 130° + 110° - 180°$$

$$\angle QRS = 60°$$

Problem 6.5.4. In Fig. (6.5.4), if $AB \parallel CD$, $\angle APQ = 50°$ and $\angle PRD = 127°$, find x and y.

6. LINES AND ANGLES

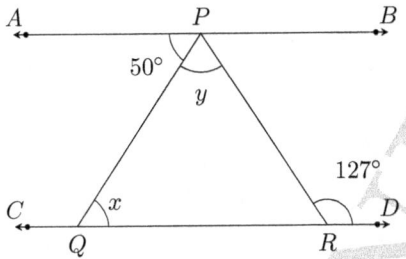

Figure 6.5.4

Solution : In Fig. (6.5.4), The line $AB \parallel CD$ and PQ is transversal line segment, so

$$\angle APQ = \angle PQR$$

(Alternate angle axiom)

$$x = 50°$$

Next $AB \parallel CD$ and PR is transversa;l line. So,

$$\angle APR = \angle PRD$$

$$\angle APQ + \angle QPR = 127°$$

$$y = 127° - 50° = 77°$$

Problem 6.5.5. In Fig. (6.5.5), PQ and RS are two mirrors placed parallel to each other. An incident ray AB strikes the mirror PQ at B, the reflected ray mobes along the path BC and strikes the mirror RS at C and again reflects back along CD. Prove that $AB \parallel CD$

6. LINES AND ANGLES

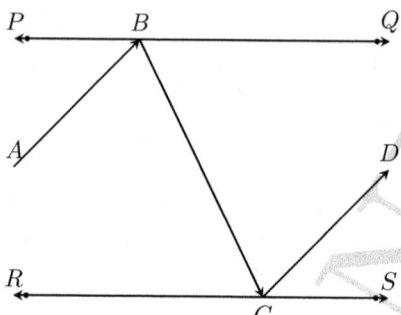

Figure 6.5.5

Proof. Construct the mormal BE and CF at the points B and C in Fig. (6.5.5). Now in Fig. (6.5.6), AB is incident ray and BC is reflected ray at point B on mirror PQ. So,

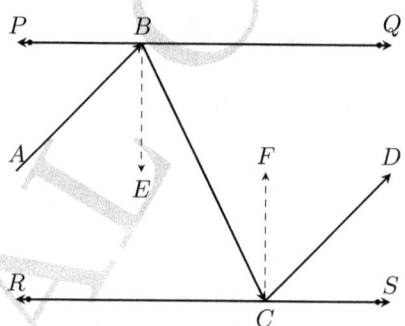

Figure 6.5.6

(6.5.1) $$\angle ABE = \angle EBC$$

(By law of reflection.)

6. LINES AND ANGLES

BC is incident ray and CD is reflected ray at point C on mirror RS. So,

(6.5.2) $$\angle BCF = \angle FCD$$

(By law of reflection.)

The normal $BE \parallel FC$, because they are mornal on parallel line PQ and RS. The line BC is transversal line.

Therefore,

(6.5.3) $$\angle EBC = \angle BCF$$

(alternate angle axiom) You get

$$2\angle EBC = 2\angle BCF$$

From equation (6.5.1), (6.5.2) and (6.5.3),

(6.5.4) $$\angle ABC = \angle BCD$$

Hence,

$$AB \parallel CD.$$

(Reverse alternate angle axiom.)

7 TRIANGLES

7.1 Introduction

You have studied the plane shape constructed through three intersecting lines. For this shape we use terminology ' **triangle**'. A triangle has six part three angles, three sides and three veretices. In previous classes you have been studied many relation and properties between the parts of triangle but in the present chapter some of them.

7.2 Congruence of Triangles

In your daily life you observed that any page of youe book, all mobiles of same company and model, two ATM cards issued by the same bank are identical. These figure are equal in all aspects or these figures are equal in shape and sizes. Such figuers are called **congruent figures**.

Now, if draw two circles of the same radius and place one on the other. You will see that they cover each other completely and you call them as congruent circles. You think if the sides and angles of one triangle are equal to the corresponding sides and angles of the other triangle, then their crresponding vertices will be same positions. Hence the all parts of one triangle are equal to the all correponding part of the other triangle. These two triangles are

7. TRIANGLES

'congruent triangle'.

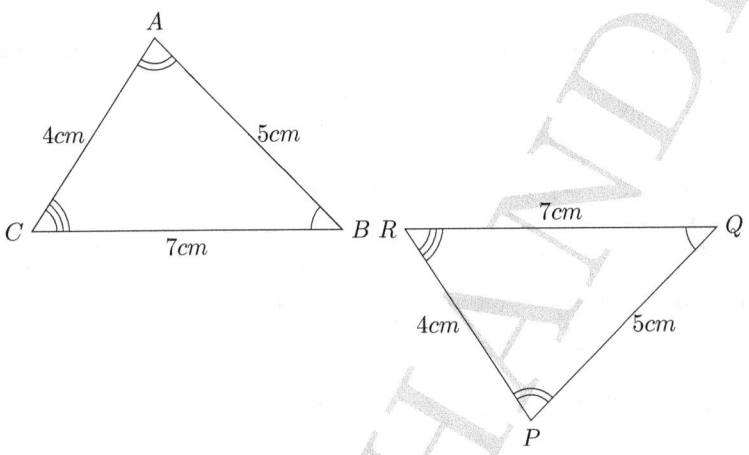

Figure 7.2.1

In the given Fig. (7.2.1), $\triangle ABC$ and $\triangle PQR$, PQ covers AB, QR covers BC, RP covers CA; $\angle Q$ covers $\angle B$, $\angle P$ covers $\angle A$, $\angle R$ covers $\angle C$, P correspond to A, Q correspond to B, R correspond to C, that is $P \Leftrightarrow A, Q \Leftrightarrow B, R \Leftrightarrow C$.

Hence $\triangle ABC$ and $\triangle PQR$ congruent, symbolicaly it is writen as

$$\triangle ABC \cong \triangle PQR$$

Remark 7.2.1. $\triangle ABC \cong \triangle PQR \not\Rightarrow \triangle ABC \cong \triangle QRP$

In congruent triangles corresponding parts are equal and it is written in short " CPCT" for *corresponding part of congruent triangles.*

7. TRIANGLES

7.3 Criteria for Congruence of Triangles

The criteria for congruence of triangles are that conditions through them unique triangle can be form.

Definition 7.3.1 (Axiom SAS congruence rule). Two triangles are congruent if two sides and the included angle of one triangle are equal to the two sides and the included angle of the other triangle.

Example 7.3.2. In Fig. (7.3.1), $OA = OB$ and $OD = OC$. Show that '

(i) $\triangle AOD \cong \triangle BOC$

(ii) $AD \parallel BC$

Solution : (i) In the Fig. (7.3.1), consider $\triangle AOD$ and $\triangle BOC$, given that,

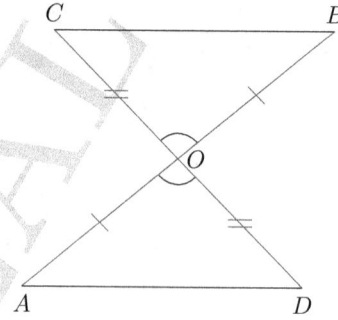

Figure 7.3.1

$OA = OB \quad OD = OC.$

7. TRIANGLES

$\angle BOC$ and $\angle AOD$ are vertically opposite angles, you get

$$\angle BOC = \angle AOD$$

So,

$$\triangle BOC \cong \triangle AOD$$

(SAS congruence axiom)

(ii) In congruent $\triangle AOD$ and $\triangle BOC$, the other corresponding parts are also equal.

So,

$$\angle OCB = \angle ODA$$

These angle are alternate angle for line segment AD and BC. So, by reverse alternate angle axiom,

$$AD \parallel BC.$$

Example 7.3.3. AB is a line segment and line l is its perpendicular bisector. If a point P lies on l, show that P is equidistant from A and B.

Solution : In the Fig. (7.3.2), line $l \perp AB$ and bisected at point C. So, C is the mid-point of AB.

Consider $\triangle PAC$ and $\triangle PCB$

$$AC = BC$$

7. TRIANGLES

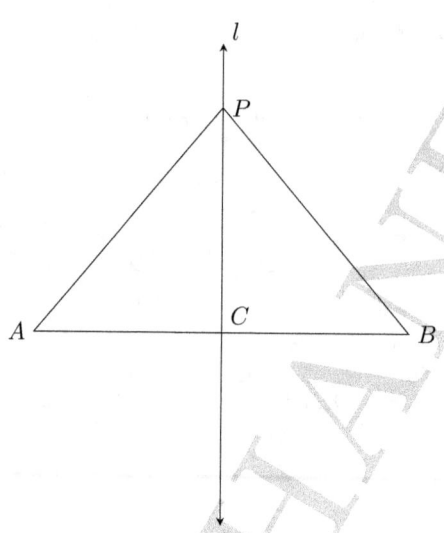

Figure 7.3.2

(C is the mid-point of AB)

$$\angle PCA = \angle PCB = 90°$$

(line l is perpendicular bisector of AB.)

$$PC = PC$$

(Common sides of triangle.)

So,

$$\triangle PAC \cong \triangle PCB$$

(SAS axiom)

Hence, $PA = PB$, because they are corresponding sides of congruent triangles.

7. TRIANGLES

Remark 7.3.4. It is very important that the equal angles are included between the pair of equal sides. So, SAS congruence axiom holds but ASS or SSA not axiom.

Theorem 7.3.5 (ASA congruence rule): *Two triangles are congruent if two angles and the included side of one triangle are equal to two angle and included side of other triangle.*

Proof. We are given two $\triangle ABC$ and $\triangle DEF$ in which

$$\angle B = \angle E, \quad \angle C = \angle F$$

and

$$BC = EF$$

We need to prove that

$$\triangle ABC \cong \triangle DEF$$

For proving the congruence of the two triangles see the cases aries.

Case (i) : Consider the Fig. (7.3.3). Let $AB = DE$

Think as,

$AB = DE$ (Assumed), $\angle B = \angle E$ (Given),

$BC = EF$ (Given), So,

$$\triangle ABC \cong \triangle DEF$$

(SAS axiom)

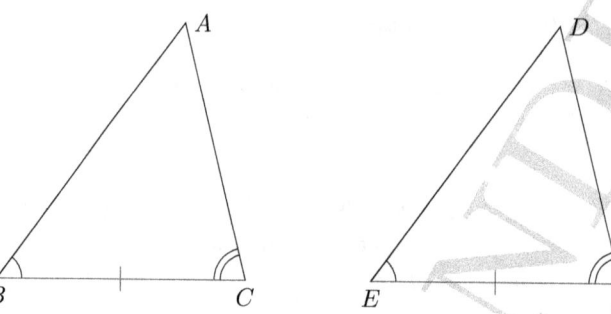

Figure 7.3.3

Case (ii) : Let if possible $AB < DE$. Then you can take a point P on DE such that $AB = EP$. Now consider $\triangle ABC$ and $\triangle PEF$ as in Fig. (7.3.4),

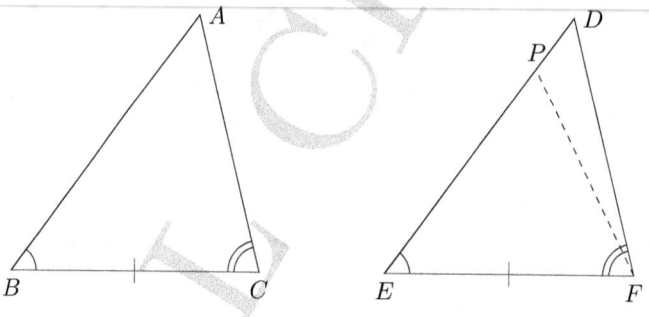

Figure 7.3.4

Think that in $\triangle ABC$ and $\triangle PEF$,

$AB = PE$ \qquad (By construction), \quad $\angle B = \angle E$ \qquad (Given),

$BC = EF$ \qquad (Given). so, on the above fact,

$\triangle ABC \cong \triangle PEF$ (By SAS axiom) When two triangles are congruent then

7. TRIANGLES

corresponding parts will be equal.

So,
$$\angle ACB = \angle PFE$$

But we are given that
$$\angle ACB = \angle DFE$$

So,
$$\angle PFE = \angle DFE$$

This possible only if P coincides with D. It means $AB = ED$. Hence $\triangle ABC \cong \triangle DEF$.

Case (iii) : if $AB > DE$, We can take a point H on AB such that $BH = ED$. So, similar logic of case (ii), we can conclude that $AB = DE$ and so, $\triangle ABC \cong \triangle DEF$. $\square$

Example 7.3.6. Line-segment AB is parralel to another line-segment CD. O is the mid-point of AD. show that (i) $\triangle AOB \cong \triangle DOC$ (ii) O is also the mid-point of BC.

Solution : (i) In Fig.(7.3.5), consider $\triangle AOB$ and $\triangle DOC$,

$\angle AOB = \angle DOC$ \qquad (Vertically opposite angles)

7. TRIANGLES

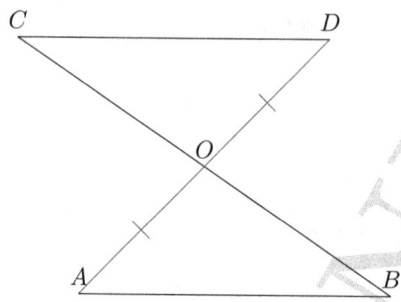

Figure 7.3.5

$\angle BAO = \angle ODC$ \quad (alternate angle axiom)

$OA = OD$ \quad (Given)

therefore, $\triangle AOB \cong \triangle DOC$ \quad (ASA rule).

Case (ii): We have $\triangle AOB \cong \triangle DOC$

So, $OB = OC$ (By CPCT), Hence O is the mid-point of BC.

7.4 Solution of Exercise 7.1

Problem 7.4.1. In Fig. (7.4.1), quadrilateral ABCD, $AC = AD$ and AB bisects $\angle A$ Show that $\triangle ABC \cong \triangle ABD$. What can you say about BC and BD ?

Solution : In Fig. (7.4.1), the quadrilateral ABCD have sides, $AC = AD$ and $\angle BAC = \angle DAB$.

7. TRIANGLES

Consider $\triangle ABC$ and $\triangle ABD$

We are given that, $\angle BAC = \angle DAB$ (Given that line AB bisects $\angle A$),

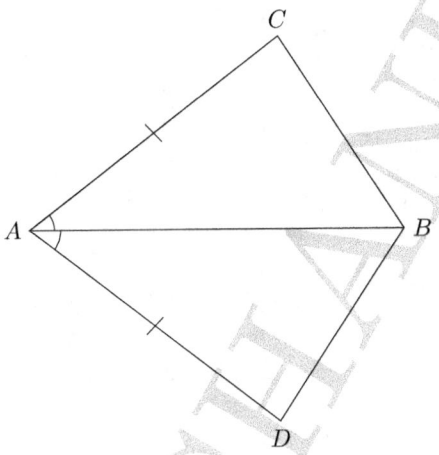

Figure 7.4.1

$AC = AD$ (Given), $AB = AB$ (Given common side).

Hence
$$\triangle ABC \cong \triangle ABD$$

(SAS congruence rule)

$$BC = BD$$

(By CPCT)

7. TRIANGLES

Problem 7.4.2. In Fig. (7.4.2), ABCD is a quadrilateral in which $AD = BC$ and $\angle DAB = \angle CBA$. Prove that,

i $\triangle ABD \cong \triangle BAC$

ii $BD = AC$

iii $\angle ABD = \angle BAC$

Solution : In Fig. (7.4.2), cosider $\triangle BAD$ and $\triangle CBA$

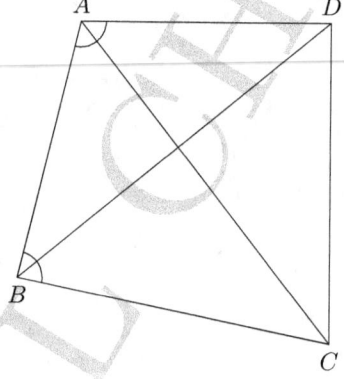

Figure 7.4.2

$$AD = BC \qquad (Given)$$

$$AB = AB \qquad (Common\ side)$$

$$\angle DAB = \angle CBA \qquad (Given)$$

7. TRIANGLES

Hence

$$\triangle ABD \cong \triangle BAC \qquad (By\ SAS\ axiom).$$

So, $BD = AC$ and $\angle ABD = \angle BAC$, (By CPCT).

Problem 7.4.3. In Fig. (7.4.3), AD and BC are equal perpendicualrs to a line segment AB. Show that CD bisects AB.

Solution : In Fig. (7.4.3), the line $BC \perp BA$ and $DA \perp AB$. Therefore, line $BC \parallel AD$. Consider $\triangle BCO$ and $\triangle ADO$.

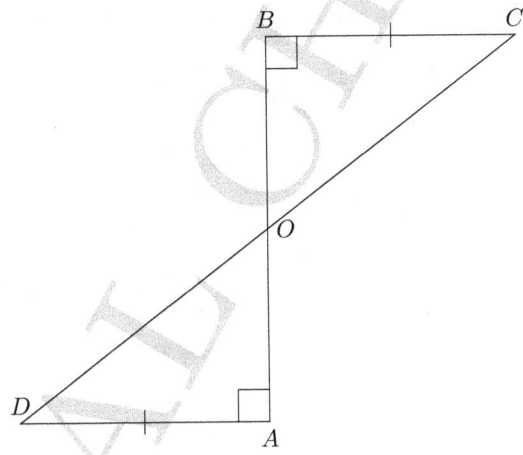

Figure 7.4.3

$$\angle CBO = \angle DAO = 90°$$

$$\angle OCB = \angle ODA$$

7. TRIANGLES

(Alternate angle axiom)

$$CB = DA$$

(Given)

Hence,

$$\triangle BCO \cong \triangle ADO$$

Now, we get $OB = OA$ (By CPCT).

We get conclusion that CD bisects AB.

Problem 7.4.4. l and m are two parallel lines intersected by another pair of parallel lines p and q. Show that $\triangle ABC \cong \triangle CDA$.

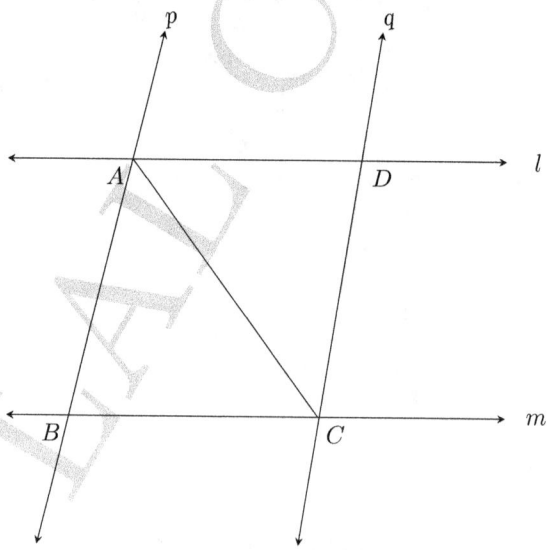

Figure 7.4.4

7. TRIANGLES

In the Fig. (7.4.4), the line-segment AC intersects the parallel lines l and m. So, $\angle DAC = \angle ACB$. The line-segment AC intersects to parallel lines p and q. So, $\angle DCA = \angle CAB$.

Consider the $\triangle ADC$ and $\triangle CAB$,

$\angle DAC = \angle ACB$ \qquad Alternate angle axiom

$\angle DCA = \angle CAB$ \qquad Alternate angle axiom

The side AC is common in bothe $\triangle ADC$ and $\triangle CAB$. So,

$$\triangle ABC \cong \triangle CDA.$$

(by ASA rule)

Problem 7.4.5. line l is the bisector of an angle $\angle A$ and B is any point on l. BP and BQ are perpendiculars from B to the arms of $\angle A$ in Fig. (7.4.5). Show that :

i $\triangle APB \cong \triangle AQB$

ii $BP = BQ$ or B is equidistant from the arms of $\angle A$.

Solution : In Fig. (7.4.5), line l is angle bisector of $\angle A$ and $QA \perp QB$ and $AP \perp PB$.

7. TRIANGLES

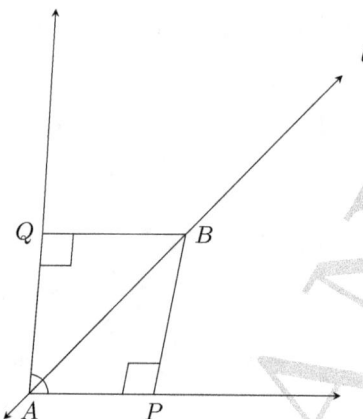

Figure 7.4.5

In $\triangle ABQ$ and $\triangle ABP$.

$$\angle BAQ = \angle BAP$$

(Line l is angle bisector of $angle QAP$)

$$\angle BQA = \angle BPA, \quad AQ \perp QB, \quad AP \perp BP$$

The line-segment AP is common side. Hence

$$\triangle ABQ \cong \triangle ABP$$

(By AAS rule)

So, $BQ = BP$ \quad (By CPCT).

Problem 7.4.6. In Fig. (7.4.6), $AC = AE, AB = AD$ and $\angle BAD = \angle EAC$. Show that $BC = DE$.

7. TRIANGLES

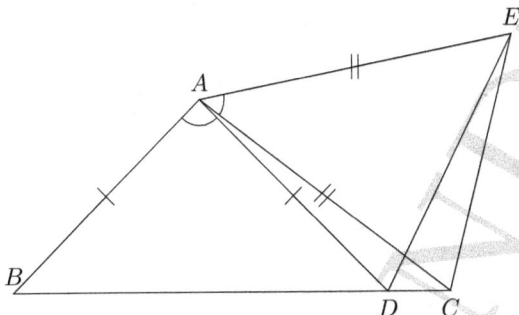

Figure 7.4.6

Solution : In Fig. (7.4.6), consider $\triangle ABC$ and $\triangle ADE$

$$AB = AD \qquad (Given)$$

$$AC = AE \qquad (Given)$$

$$\angle BAD = \angle EAC \qquad (Given)$$

So,

$$\triangle ABC \cong \triangle ADE$$

(by SAS axiom)

Therefore, $BC = DE$, (CPCT)

Problem 7.4.7. AB is a line segment and P is its mid-point. D and E are points on the same side of AB such that $\angle BAD = \angle ABE$ and $\angle EPA = \angle DPB$ according Fig. (7.4.7). Show that,

7. TRIANGLES

i $\triangle DAP \cong \triangle EBP$

ii $AD = BE$

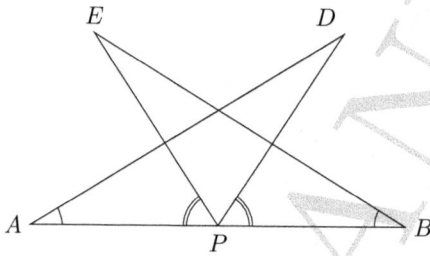

Figure 7.4.7

Solution : In Fig. (7.4.7), point P is mid point of line segment AB, so $AP = BP$. Now in $triangle DAP$ and $\triangle EBP$,

$$\angle BAD = \angle ABE \qquad (Given)$$

$$\angle EPA = \angle DPB \qquad (Given)$$

$$PA = PB$$

Therefore,

$$\triangle DAP \cong \triangle EBP$$

According ASA rule.

Hence,

$$AD = BE \qquad (CPCT)$$

7. TRIANGLES

Problem 7.4.8. In right angle $\triangle ABC$, right angled at C, M is the mid-point of hypotenuse AB. C is joined to M and produced to a point D such that $DM = CM$. Point D is joined to B (as in Fig. (7.4.8)). Show that:

i $\triangle AMC \cong \triangle BMD$

ii $\triangle DBC$ is a right angle.

iii $\triangle DBC \cong \triangle ACB$

iv $CM = \frac{1}{2}AB$

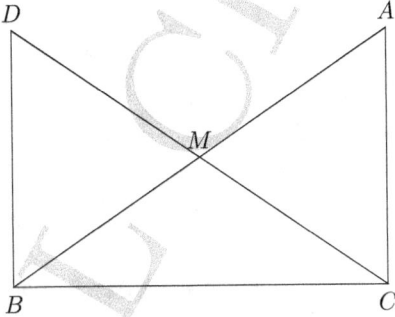

Figure 7.4.8

Solution : In Fig. (7.4.8), M is the mid-point of AB, and C is a right angle and $DM = CM$. Consider in $\triangle AMC$ and $\triangle BMD$

$$AM = BM \qquad (Given)$$

7. TRIANGLES

$$DM = MC \quad (Given)$$

$$\angle AMC = \angle BMD \quad (Vertically\ opposite\ angle)$$

Hence

$$\triangle AMC \cong \triangle BMD \quad (SAS\ axiomn)$$

Now $AC = BD$ and $\angle ACB = \angle DBC$ (By CPCT)

But $\angle BCA = 90°$. So,

$$\angle DBC = 90°$$

Consider in $\triangle DBC, \triangle ACB$

$$AC = BD \quad (Given)$$

$$BC = BC \quad (Common)$$

$$\angle ACB = \angle DBC = 90° \quad (Given)$$

Hence

$$\triangle DBC \cong \triangle ACB \quad (SAS\ axiom)$$

Therefore, $AB = DC$ (By CPCT).

$$DC = 2MC = AB \implies MC = \frac{1}{2}AB.$$

7. TRIANGLES

7.5 Some Properties os a Triangle

In the section (7.4), we have studied two criteria for congruence of a triangles. In these two criteria, one is axiom and other is theorem. Now we find out results apply this criteria of congruence of triangles.

Theorem 7.5.1: *Angles opposite to equal sides of an isosceles triangle are equal.*

Proof. We are given an isosceles triangle ABC in which $AB = AC$. We need to prove that $\angle B = \angle C$.

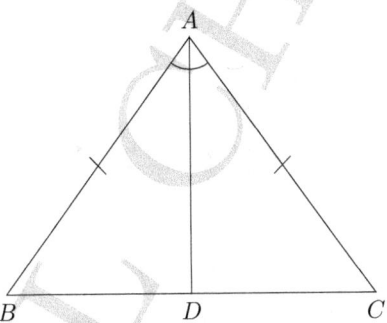

Figure 7.5.1

Let us draw angle bisector of $\angle A$ which meets at point D, on BC. So, $\angle BAD = \angle CAD$.

Consider in $\triangle BAD$ and $\triangle CAD$

$$AB = AC \qquad (Given)$$

7. TRIANGLES

$$\angle BAD = \angle CAD \qquad (By\ construction)$$

$$AD = AD \qquad (Commmon)$$

So,

$$\triangle BAD \cong \triangle CAD \qquad (By\ SAS\ axiom)$$

so, $\angle ABD = \angle ACD$, since they are corresponding angles of congruent triangles. So,

$$\angle B = \angle C.$$

$\square$

Theorem 7.5.2: *Angle opposite to equal sides of an isosceles triangle are equal.*

Proof. We are given an isosceles triangle ABC in which $\angle B = \angle C$. We need to prove that $AB = AC$.

Let us draw angle bisector of $\angle A$ which meets at point D, on BC. So, $\angle BAD = \angle CAD$.

Consider in $\triangle BAD$ and $\triangle CAD$

$$\angle B = \angle C \qquad (By\ construction)$$

$$\angle BAD = \angle CAD \qquad (Given)$$

7. TRIANGLES

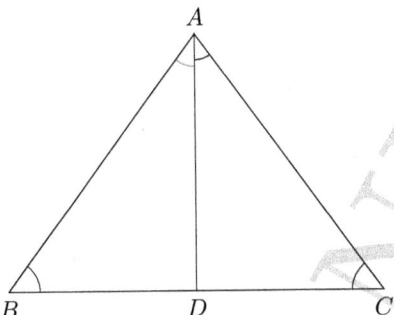

Figure 7.5.2

$$AD = AD \quad (Commmon)$$

So,
$$\triangle BAD \cong \triangle CAD \quad (By\ AAS\ rule)$$

so,
$$AB = AC.$$

, since they are corresponding sides of congruent triangles. So,

$$AB = AC.$$

□

Example 7.5.3. In $\triangle ABC$, the bisector AD of $\angle A$ is perpendicular to side BC (given Fig. 7.5.3). Show that $ABAC$ and $\triangle ABC$ is isosceles.

7. TRIANGLES

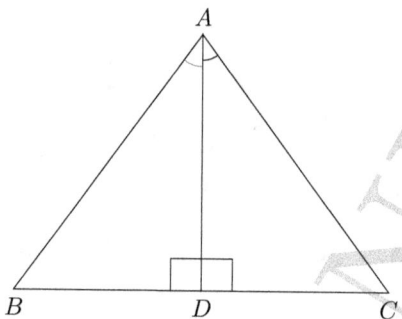

Figure 7.5.3

Solution :

Consider in $\triangle ADB$ and $\triangle ADC$,

$$\angle DAB = \angle DAC \qquad (Given)$$

$$\angle ADB = \angle ADC = 90° \qquad (Given)$$

$$DA = DA \qquad (Common\ sides)$$

So,

$$\triangle ABD \cong \triangle ADC \qquad (By\ ASA\ rule)$$

Hence,

$$AB = AC \qquad (CPCT)$$

Therefore, $\triangle ABC$ is an isosceles triangle.

Example 7.5.4. E and F are respectively the mid-points of equal sides AB and AC of $\triangle ABC$ Given in Fig, (7.5.4). Show that $BF = CE$.

7. TRIANGLES

Solution : In Fig. (7.5.4), $\triangle ABC$, the equal sides AB and AC have mid-points E and F. So, $\angle ABC = \angle ACB$, which means $\angle EBC = \angle FCB$.

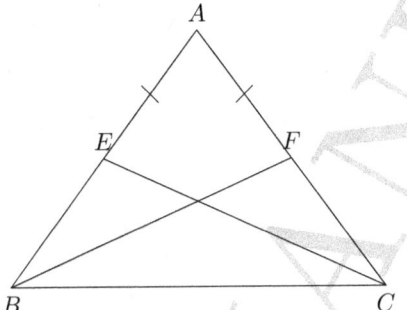

Figure 7.5.4

Now, in $\triangle BEC$ and $\triangle BFC$,

$\angle EBC = \angle FCB$ (Angle of isosceles triangle)

$EB = FC$ (Half side of isosceles triangle)

$BC = CB$ (Common side)

So,

$\triangle BEF \triangle CFB$ (SAS axiom)

Therefore,

$EC = BF$ (CPCT)

Example 7.5.5. In Fig. (7.5.5), an isosceles $\triangle ABC$ with $AB = AC$, D and E are points on BC such that $BE = CD$. Show that $AD = AE$.

7. TRIANGLES

Solution : In $\triangle ABE$ and $\triangle ACD$

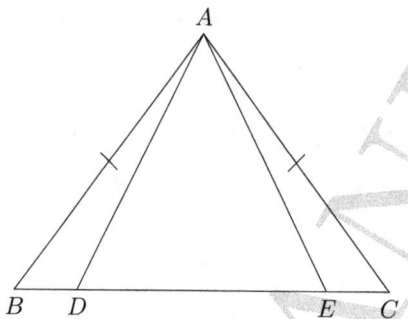

Figure 7.5.5

$$AB = AC \qquad (Given)$$

$$BE = CD \qquad (Given)$$

$$\angle B = \angle C \qquad (Angle\ opposite\ to\ equal\ sides)$$

So,
$$\triangle ABE \cong \triangle ACD \qquad (SAS, axiom)$$

Hence,
$$AE = AD \qquad (CPCT)$$

7. TRIANGLES

7.6 Solution of Exercise 7.2

Problem 7.6.1. In an isosceles $\triangle ABC$, with $AB = AC$, the bisector of $\angle A$ and $\angle C$ intersect each at O. Join A to O. Show that : (i) $OB = OC$ (ii) AO bisect $\angle A$.

Solution : In Fig. (7.6.1), take $\triangle AEB$ and $\triangle ADC$

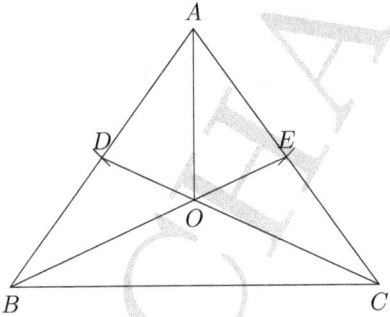

Figure 7.6.1

$\angle ABE = \angle ACD$ (half angle of opposite of equal sides)

$AB = AC$ (Given)

$\angle A = \angle A$ (Common)

So,

$\triangle ABE \cong \triangle ACD$ (ASA, rule)

7. TRIANGLES

Hence,
$$AD = AE \quad (CPCT)$$

Now, in $\triangle DOB$ and $\triangle EOC$

$$DB = EC \quad (Because\ AB = AC)$$

$$\angle DBO = \angle ECO \quad (\angle B = \angle C)$$

$$\angle DOB = \angle EOC \quad (Vertically\ opposite\ angle)$$

Hence,
$$\triangle DOB \cong \triangle EOC$$

$$OB = OC \quad (CPCT)$$

and
$$OD = OE \quad (CPCT)$$

Now, in $\triangle AOD$ and $\triangle AOE$,

$$AD = AE \quad (because\ AB = AC)$$

$$OD = OE \quad (\triangle DOB \cong \triangle EOC)$$

$$\angle ADO = \angle AEO \quad (\triangle ABE \cong \triangle ACD)$$

Therefore,
$$\triangle AOE \cong \triangle AOD$$

7. TRIANGLES

$$\angle DAO = \angle OAE \qquad (CPCT)$$

Hence, OA is bisector of $\angle A$

Problem 7.6.2. In Fig. (7.6.2), $\triangle ABC$, AD is perpendicular bisector of BC. Show that $\triangle ABC$ is an isosceles triangle in which sides $AB = AC$.

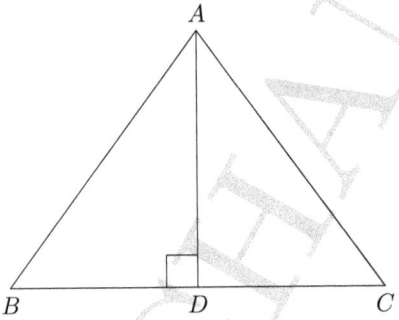

Figure 7.6.2

Solution : In Fig. (7.6.2), AD is perpendicular bisector of BC. So, $BD = DC$.

Now, in $\triangle ABD$ and $\triangle ADC$

$$\angle ADB = \angle ADC = 90° \qquad (Linear\ pair\ angle)$$

$$DB = DC \qquad (AD\ bisect\ BC)$$

$$AD = AD \qquad (Common)$$

So,

$$\triangle ADB \cong \triangle ADC \qquad (SAS\ axiom)$$

7. TRIANGLES

Hence,

$$AB = AC \qquad (CPCT)$$

Therefore, $\triangle ABC$ is isosceles triangle.

Problem 7.6.3. In Fig. (6.7.3), ABC is an isosceles triangle in which altitudes BE and CF are drawn to equal sides AB and AC respectively. Show that these altitedes are equal.

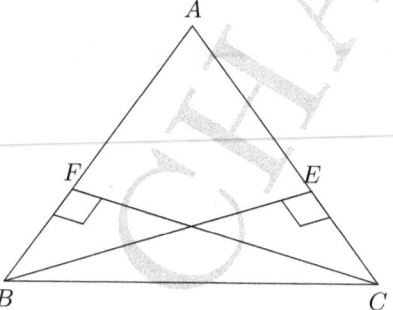

Figure 7.6.3

Solution : In Fig. (7.6.3), $\triangle ABC$ is isosceles triangle which side $AB = AC$.

Cosider in $\triangle BFC$ and $\triangle CEB$,

$$\angle FBC = \angle ECB \qquad (Opposite\ angle\ of\ equal\ sides)$$

$$\angle BFC = \angle CEB = 90° \qquad (BE \perp AC,\ CF \perp AB)$$

$$BC = BC \qquad (Common)$$

7. TRIANGLES

So,
$$\triangle BFC \cong \triangle CEB \quad (AAS\ rule)$$

Hence,
$$CFEB \quad (CPCT)$$

Problem 7.6.4. ABC is a triangle in which altitude BE and CF to sides AC and AB are equal as in Fig. (7.6.4). Show that:

i $\triangle ABE = \triangle ACF$

ii $AB = AC$, i.e ABC is an isosceles triangle.

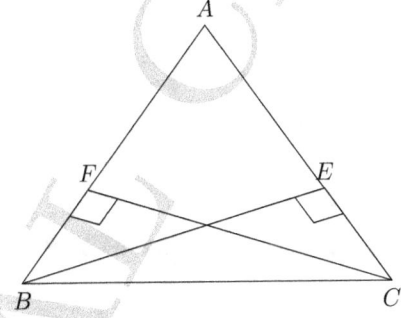

Figure 7.6.4

Solution : Cosider in $\triangle ABE$ and $\triangle ACF$,

$$\angle BFC = \angle CEB = 90° \quad (BE \perp AC,\ CF \perp AB)$$

$$BE = FC \quad (Given)$$

$$\angle A = \angle A \qquad (Common)$$

So,
$$\triangle ABE \cong \triangle ACF \qquad (AAS\ Rule)$$

Hence,
$$AB = AC \qquad (CPCT)$$

Problem 7.6.5. ABC and DBC are two isosceles triangles on the same base BC as in Fig. (7.6.5). Show that $\angle ABD = \angle ACD$.

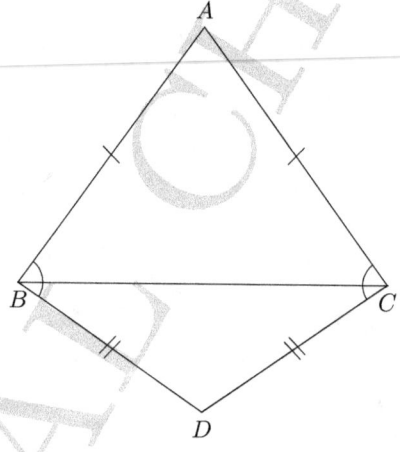

Figure 7.6.5

Now, in $\triangle ABC$ the sies AB and AC are equal. So, $\angle ABC$ and $\angle ACB$ are equal angles.

(7.6.1) $$\angle ABC = \angle ACB$$

7. TRIANGLES

Similaraly, in $\triangle BDC$,

(7.6.2) $\qquad\qquad \angle CBD = \angle BCD$

From, equation (7.6.1) and (6.7.2),

(7.6.3) $\qquad\qquad \angle ABC + \angle CBD = \angle ACB + \angle BCD$

Therefore
$$\angle ABD = \angle ACD$$

Problem 7.6.6. $\triangle ABC$ is an isosceles triangle in which $AB = AC$. Side ab produced to D such that $AD = AB$ as in Fig. (7.6.6). Show that $\angle BCD$ is a right angle.

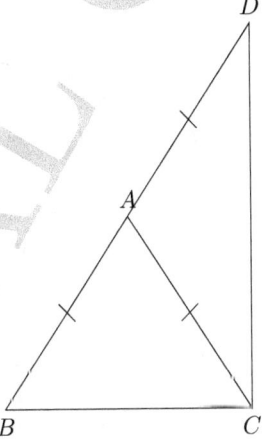

Figure 7.6.6

7. TRIANGLES

In Fig. (7.6.6), the $\triangle ABC$ and $\triangle ADC$ are isosceles triangle with $AB = AC$ and $AC = AD$ respectively. So,

$$\angle ABC = \angle ACB \qquad (Angle\ of\ isosceles\ triangle)$$

and

$$\angle ADC = \angle ACD \qquad (Angle\ of\ isosceles\ triangle)$$

$$\angle CAD = \angle ABC + \angle ACB \qquad (Exterior\ angle)$$

So,

$$\angle CAD = 2\angle ACB$$

Similarly,

$$\angle BAC = 2\angle ACD$$

$$\angle CAD + \angle BAC = 180° \qquad (Linear\ pair\ angle)$$

$$2\angle ACB + 2\angle ACD = 180°$$

$$2(\angle ACB + \angle ACD) = 180°$$

Hence,

$$\angle BCD = 90°$$

Problem 7.6.7. ABC is a right angled triangle in which $\angle A = 90°$ and $AB = AC$. Find $\angle B$ and $\angle C$.

7. TRIANGLES

Solution : In Fig.(7.6.7), ABC is right angle triangle with $\angle A = 90°$ and $AB = AC$.

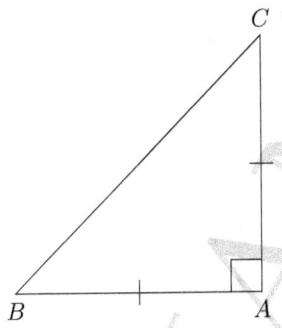

Figure 7.6.7

The $\triangle ABC$ is an isosceles triangle. Hence,

$$\angle B = \angle C \qquad (opposite\ angle\ of\ equal\ side)$$

$$\angle A + \angle B + \angle C = 180° \qquad (Triangle\ properties)$$

$$\angle B + \angle C = 90°$$

$$\angle B = 45° = \angle C \qquad (\angle B = \angle C)$$

Problem 7.6.8. Show that the angles of an equilateral triangle are 60° each.

Solution : In Fig. (7.6.8), $\triangle ABC$ is an equilateral triangle. Then it is isosceles triangle also.

Now, in $\triangle ABC$, $AB = AC$, So,

$$\angle C = \angle B \qquad (Opposite\ angle\ of\ equal\ side)$$

7. TRIANGLES

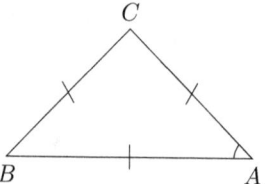

Figure 7.6.8

Similaraly,

$$\angle B = \angle C \qquad (Opposite\ angle\ of\ equal\ side)$$

Therefore,

$$\angle B = \angle C = \angle A$$

$$\angle B + \angle C + \angle A = 180° \qquad (triangle\ property)$$

Hence,

$$\angle B = \angle C = \angle A = 60°.$$

7.7 Some Other criteria for Congruence of Triangle

We have seen in section (7.5) and (7.3) the quality of congruence triangle but in these section discussion we can not derived conclusion, if three angle of a triangle are equal to three angle of other triangle. What you say if three sides of a triangle are equal to three sides of other triangle.

www.ingramcontent.com/pod-product-compliance
Lightning Source LLC
Chambersburg PA
CBHW052154220526
45471CB00004B/1670